I0018392

Creating Data Stories with Tableau Public

Illustrate your data in a more interactive and interesting way using Tableau Public

Ashley Ohmann

Matt Floyd

BIRMINGHAM - MUMBAI

Creating Data Stories with Tableau Public

Copyright © 2015 Packt Publishing

All rights reserved. No part of this book may be reproduced, stored in a retrieval system, or transmitted in any form or by any means, without the prior written permission of the publisher, except in the case of brief quotations embedded in critical articles or reviews.

Every effort has been made in the preparation of this book to ensure the accuracy of the information presented. However, the information contained in this book is sold without warranty, either express or implied. Neither the authors, nor Packt Publishing, and its dealers and distributors will be held liable for any damages caused or alleged to be caused directly or indirectly by this book.

Packt Publishing has endeavored to provide trademark information about all of the companies and products mentioned in this book by the appropriate use of capitals. However, Packt Publishing cannot guarantee the accuracy of this information.

First published: November 2015

Production reference: 1251115

Published by Packt Publishing Ltd.
Livery Place
35 Livery Street
Birmingham B3 2PB, UK.

ISBN 978-1-84969-476-6

www.packtpub.com

Credits

Authors
Ashley Ohmann

Matt Floyd

Reviewer
Joshua N. Milligan

Acquisition Editor
Meeta Rajani

Content Development Editor
Pooja Mhapsekar

Technical Editor
Vishal Mewada

Copy Editor
Vedangi Narvekar

Project Coordinator
Nidhi Joshi

Proofreader
Safis Editing

Indexer
Hemangini Bari

Graphics
Kirk D'Penha

Production Coordinator
Melwyn Dsa

Cover Work
Melwyn Dsa

About the Author

Ashley Ohmann started her career in technology as a Flash and HTML developer at the Emory University School of Medicine while studying Classics as an undergraduate at Emory University. After learning how to write SQL to help create a fraud detection system for a client, she pursued information management and data analytics as a vocation. While working for a multinational manufacturing company, she was asked to evaluate Tableau Desktop and Tableau Server; consequently, her team became one of the first to implement the suite of tools for their enterprise.

Ashley's career with Tableau's products has included work as a consultant, trainer, and a professional services practice director. She is a frequent contributor to the San Francisco and Phoenix Tableau User Groups.

A native of Highlands, NC and Atlanta, GA, Ashley is a proud alumna of Rabun Gap – Nacoochee School. She also studied German and Chemistry at Mount Holyoke College before graduating from Emory. Ashley's roots go back to south Georgia; she grew up listening to the stories of her large extended family, which inspired her to spend her career helping other people learn how to tell their own stories using a variety of media. Currently, she lives in the Pacific Northwest with her family, where they enjoy skiing, the beauty of God's great creation, and practicing permaculture on their 10 acre farm.

Acknowledgments

There are many amazing people who have encouraged and inspired me on this journey with Tableau, which began in early 2011. My husband and son have sacrificed countless weekends so that I could write, and they have been amazingly supportive. I am very blessed to have them.

In 2011, I returned to work from maternity leave, and my manager, *Sandeep Sivadas*, asked me to use Gartner's Magic Quadrant to evaluate which BI tool would best fit our needs. After many downloads, tests, and phone calls with *John Jensen*, *Mike Kravec*, and *Jeremy Walsh*, all from Tableau, we selected Tableau Desktop and Server, and with Sandeep's leadership, we truly changed our organization with it (thank you, Sandeep!). Many thanks to JJ, Mike, and Jeremy for changing the courses of thousands of their customers across the country; each of you is amazing and has a special place in my heart.

There are many Tableau User Group leaders who have created and sustained communities of users and encouraged me to present and later become a leader. Thanks to the crew at ATUG, John and Susana at SFBATUG, Michael at PHXTUG, and Lauren Rogers, who has expanded the TUG network so phenomenally well.

None of this would have been possible without the people at Tableau who design and create the products—and the partner network—with a constant focus on helping other people see and understand their data. Thank you all for executing a world-changing vision.

Lastly, thanks to my parents, who encouraged me to start writing 20 years ago and always made sure I had the tools and encouragement to let my light shine.

About the Author

Matt Floyd has worked in the software industry since 2000 and has held career roles from project management to technical writing and business intelligence analysis. His career has spanned many industries, including environment, healthcare, pharmaceuticals, and insurance.

Matt's hands-on experience with Tableau started in 2008 after evaluating alternatives to reporting and analytical software used by his clients. Since then, he has been a technical writer, implementation engineer, consultant, developer, and analyst in BI projects. His passion for Tableau stems from his fascination of discovery through data and the art, science, and power of data visualization. He is currently interested in text mining and the combination of that data with powerful visualizations that tell fascinating stories. He and his family live in metro Atlanta, and when not stuck in traffic, he sometimes offers musings on his blog covering various visualization topics at http://floydmatt.com/.

Thank you to my wife, *Beth*, for her love and patience as I worked on this book as well as my daughters, *Audrey* and *Hope*, for being reliably funny and understanding. Thank you to my co-author, *Ashley*, for the technical chops and experience to make this an outstanding the book. To my employer, thank you for finally getting Tableau (great job, Jim K, for making it happen!), and thanks to all the Matts who make this world awesome.

About the Reviewer

Joshua N. Milligan has been a consultant with Teknion Data Solutions since 2004 and he currently serves as a team lead and project manager. With a strong background in software development and custom .NET solutions, he brings a blend of analytical and creative thinking to BI solutions, data visualization, and data storytelling. His years of consulting have given Joshua hands-on experience with all aspects of the BI development cycle, which includes data modeling, ETL, enterprise deployment, data visualization, and dashboard design. He has worked with clients in numerous industries, including finance, healthcare, marketing, government, and services.

In 2014 and again in 2015, Joshua was named Tableau Zen Master, the highest recognition of excellence from Tableau Software. As a Tableau-accredited trainer, mentor, and leader in the online Tableau community, he is passionate about helping others gain insights from their data. His work has been featured multiple times on Tableau Public's Viz of the Day and Tableau's website. He also shares frequent Tableau tips, tricks, and advice on his blog, which can be viewed by visiting `http://vizpainter.com/`. He is the author of *Learning Tableau*.

I owe a debt of gratitude to many who have mentored, guided, and taught me throughout the years. My father, Stuart, opened up the world of computer programming to me and also imparted a passion to help others. Thank you to all the individuals at Teknion Data Solutions—my colleagues with whom I have the privilege to collaborate on a daily basis, the management, and its owners, who have made an investment in our training and growth and created an exciting place to build a career. Thank you to the Tableau employees and members of the online community who have created an incredible place to learn, share, help others, and have fun. Most of all, thank you to my wonderful wife, Kara, who has supported and encouraged me all along the way.

www.PacktPub.com

Support files, eBooks, discount offers, and more

For support files and downloads related to your book, please visit www.PacktPub.com.

Did you know that Packt offers eBook versions of every book published, with PDF and ePub files available? You can upgrade to the eBook version at www.PacktPub.com and as a print book customer, you are entitled to a discount on the eBook copy. Get in touch with us at service@packtpub.com for more details.

At www.PacktPub.com, you can also read a collection of free technical articles, sign up for a range of free newsletters and receive exclusive discounts and offers on Packt books and eBooks.

https://www2.packtpub.com/books/subscription/packtlib

Do you need instant solutions to your IT questions? PacktLib is Packt's online digital book library. Here, you can search, access, and read Packt's entire library of books.

Why subscribe?

- Fully searchable across every book published by Packt
- Copy and paste, print, and bookmark content
- On demand and accessible via a web browser

Free access for Packt account holders

If you have an account with Packt at www.PacktPub.com, you can use this to access PacktLib today and view 9 entirely free books. Simply use your login credentials for immediate access.

Table of Contents

Preface

Tableau Software is on a mission to help people see and understand their data. Tableau Public, which is their free tool that allows anyone to publish interactive visualizations in the cloud, is a tremendous step toward democratizing data by providing tools to data journalists, bloggers, and hobbyists that previously would have been available only through corporate IT departments.

When we initially started off this project, our goal was to describe the features of Tableau Public so that you can create your own stories with data and then share them with the community. We also wanted to provide examples of how members of the online community have used Tableau Public to draw attention to important issues of our time. In the intervening months, many things have transpired: in addition to the important features in Tableau version 9.x, the online community has multiplied in size, about 500,000 people a day use Tableau Public, as of this writing, and the Tableau Foundation was established. Its mission is to use data to make a tangible difference in the world through the involvement of data volunteers and by granting software licenses to non-profit organizations that are dedicated to improving public health and the lives of underprivileged populations around the world.

Tableau Public is only as good as the people using it, and as we progressed on our own journeys with Tableau Public, we became engrossed in the data stories and dialogue that people have created and curated, particularly those of the Tableau foundation volunteers and others that have the express purpose of using data for good. We have attempted to provide guidelines for you on how to craft a compelling, rich story with data that enlightens others. Our examples focus on variations of the World Bank indicators, and by the end of this book, we have crafted an example dashboard that focuses on an issue that impacts everyone.

What this book covers

Chapter 1, Getting Started with Tableau Public, you will have an overview of the uses and functions of Tableau Public, understand how dashboards are used, and how to download and install Tableau Public.

Chapter 2, Tableau Public Interface Features, you will have an understanding of the various features in Tableau Public, such as cards, shelves, and ShowMe.

Chapter 3, Connecting to Data, will teach you the various ways in which source data can be formatted, and will explain some basic data modeling such as Dimensions, Measures, and joins. You will also understand more about using multiple data sources and data blending in dashboards.

Chapter 4, Visualization -Tips and Types, will present the types of visualizations available for use in Tableau Public, and how to use them with their data.

Chapter 5, Calculations, will guide you how to best use calculations, basic statistics, and predictive analytics in Tableau Public.

Chapter 6, Level of Detail and Table Calculations, will discuss how you can use Table Calculations and Level of Detail calculations to enhance the comparisons that you are making with data and also how to make them more dynamic and contextual.

Chapter 7, Dashboard Design and Styling, you will understand the basics of good dashboard design, and have an overview of data visualization best practices using Tableau Public.

Chapter 8, Filters and Actions, will explain how to build filters and actions for use in their dashboards.

Chapter 9, Publishing Your Work, will familiarize you with the various methods to embed data visualizations in blog posts, websites, and offline documents.

What you need for this book

Users only need to download the Tableau Public client. The technical specifications for Tableau Public mirror those of Tableau Desktop Personal and are listed on the Tableau website at http://www.tableau.com/products/desktop.

According to Tableau system requirements, PC users require the following minimum specifications:

- Microsoft Windows Vista SP2 or newer (32-bit and 64-bit)
- Microsoft Server 2008 R2 or newer (32-bit and 64-bit)

- Intel Pentium 4 or AMD Opteron processor or newer (SSE2 or newer required)
- 2 GB memory
- 750 megabytes minimum free disk space
- Internet Explorer 8 or newer

And Mac users need the following specifications in their system:

- iMac/MacBook computers 2009 or newer
- OS X 10.9 or newer

Who this book is for

This book is targeted at investigative journalists and bloggers with an interest in making rich and interactive data visualizations. Intermediate Tableau Public users and organizations can also use this book as a reference guide and teaching aid. Members of the media team, such as data specialists, web developers, editors, producers, and managers can also benefit from an understanding of the structure and challenges of writing an interactive and interesting data visualization using Tableau Public.

Conventions

In this book, you will find a number of text styles that distinguish between different kinds of information. Here are some examples of these styles and an explanation of their meaning.

Code words in text, database table names, folder names, filenames, file extensions, pathnames, dummy URLs, user input, and Twitter handles are shown as follows: "Gather your data sources, usually in a spreadsheet or a .csv file."

New terms and **important words** are shown in bold. Words that you see on the screen, for example, in menus or dialog boxes, appear in the text like this: "Enter your email address and click on the **Download the App** button located in the middle of the screen."

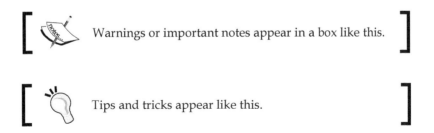

Warnings or important notes appear in a box like this.

Tips and tricks appear like this.

Reader feedback

Feedback from our readers is always welcome. Let us know what you think about this book, what you liked or disliked. Reader feedback is important for us as it helps us develop titles that you will really get the most out of.

To send us general feedback, simply e-mail feedback@packtpub.com, and mention the book's title in the subject of your message.

If there is a topic that you have expertise in and you are interested in either writing or contributing to a book, see our author guide at www.packtpub.com/authors.

Customer support

Now that you are the proud owner of a Packt book, we have a number of things to help you to get the most from your purchase.

Downloading the example code

You can download the example code files from your account at http://www.packtpub.com for all the Packt Publishing books you have purchased. If you purchased this book elsewhere, you can visit http://www.packtpub.com/support and register to have the files e-mailed directly to you.

Downloading the color images of this book

We also provide you with a PDF file that has color images of the screenshots/diagrams used in this book. The color images will help you better understand the changes in the output. You can download this file from http://www.packtpub.com/sites/default/files/downloads/3109OT_ColorImages.pdf.

Errata

Although we have taken every care to ensure the accuracy of our content, mistakes do happen. If you find a mistake in one of our books — maybe a mistake in the text or the code — we would be grateful if you could report this to us. By doing so, you can save other readers from frustration and help us improve subsequent versions of this book. If you find any errata, please report them by visiting http://www.packtpub.com/submit-errata, selecting your book, clicking on the **Errata Submission Form** link, and entering the details of your errata. Once your errata are verified, your submission will be accepted and the errata will be uploaded to our website or added to any list of existing errata under the Errata section of that title.

To view the previously submitted errata, go to `https://www.packtpub.com/books/content/support` and enter the name of the book in the search field. The required information will appear under the **Errata** section.

Piracy

Piracy of copyrighted material on the Internet is an ongoing problem across all media. At Packt, we take the protection of our copyright and licenses very seriously. If you come across any illegal copies of our works in any form on the Internet, please provide us with the location address or website name immediately so that we can pursue a remedy.

Please contact us at `copyright@packtpub.com` with a link to the suspected pirated material.

We appreciate your help in protecting our authors and our ability to bring you valuable content.

Questions

If you have a problem with any aspect of this book, you can contact us at `questions@packtpub.com`, and we will do our best to address the problem.

1
Getting Started with Tableau Public

Making sense of data is a valued service in today's world. It may be a cliché, but it's true that we are drowning in data and yet, we are thirsting for knowledge. The ability to make sense of data and the skill of using data to tell a compelling story is becoming one of the most valued capabilities in almost every field, which includes business, journalism, retail, manufacturing, medicine, and public service. Tableau Public (for more information, visit www.tableaupublic.com), which is Tableau Software's free Cloud-based data visualization client, is a powerfully transformative tool that can be used to create rich, interactive, and compelling data stories. It's a great platform if you wish to explore data through visualization. It enables your consumers to ask and answer questions that are interesting to them.

This book is written for people who are new to Tableau Public and who would like to learn how to create rich, interactive data visualizations from publicly available data sources that they then can easily share with others. Once you publish visualizations and data to Tableau Public, they are accessible to, and can be viewed and downloaded by, everyone. A typical Tableau Public data visualization contains public data sets such as sports, politics, public works, crime, census, socioeconomic metrics, and social media sentiment data. You can also create and use your own data. Many of these data sets are either readily available on the Internet, or can be accessed via a public records request or search (if they are harder to find, they can be scraped from the Internet). You can now control who can download your visualizations and data sets, which is a feature that was previously available only to paid subscribers. Tableau Public currently has a maximum data set size of 10 million rows and/or 10 GB of data.

In this chapter, we will walk through an introduction to Tableau, which includes the following topics:

- A discussion on how you can use Tableau Public to tell your data story
- Examples of organizations that use Tableau Public
- Downloading and installing the Tableau Public software
- Logging in to Tableau Public
- Creating your own Tableau Public profile
- Discovering the Tableau Public features and resources
- Having a look at the author profiles and galleries section of the Tableau website so that we can browse other authors' data visualizations (this is a great way to learn and gather ideas on how to best present data)

A Tableau Public overview

Tableau Public allows you to tell your data story and create compelling and interactive data visualizations that invite discovery and education. Tableau Public is sold at a great price—free. It allows you as a data storyteller to create and publish data visualizations without learning how to code or having special knowledge about web publishing. In fact, you can publish data sets of up to 10 million rows or 10 GB to Tableau Public in a single workbook. Tableau Public is a data discovery tool. It should not be confused with enterprise-grade business intelligence tools, such as Tableau Desktop and Server, QlikView, and Cognos Insight. Those tools integrate with corporate networks and security protocols as well as server-based data warehouses. Data visualization software is not a new thing. Businesses have used software to generate dashboards and reports for decades. The new twist comes with data discovery tools, such as Tableau Public. Journalists and bloggers who would like to augment their reporting of static text and graphics can use these data discovery tools, such as Tableau Public, to create compelling, rich data visualizations, which may consist of one or more charts, graphs, tables, and other objects that can be controlled by readers to allow for discovery.

The people who are active members of the Tableau Public community have a few primary traits in common— they are curious, generous with their knowledge and time, and enjoy conversations that relate data to the world around us. Tableau Public maintains a list of blogs of data visualization experts who use Tableau Software.

In the following screenshot, Tableau Zen Masters *Anya A'hearn* of Databrick and *Allan Walker* used data on San Francisco bike sharing to show the financial benefits of **Bay Area Bike Share**, a city-sponsored 30-minute bike sharing program, as well as a map of both the proposed expansion of the program and how far a person can actually ride a bike in half an hour.

This dashboard is featured in the Tableau Public gallery because it relates data to users clearly and concisely. It presents a great public interest story (commuting more efficiently in a notoriously congested city) and then grabs the viewer's attention with maps of current and future offerings. The second dashboard within the analysis is significant, as well. The authors described the Geographic Information Systems (**GIS**), the tools that they used to create innovative maps, as well as the methodology that went into the final product so that the users who are new to the tool can learn how to create a similar functionality for their own purposes:

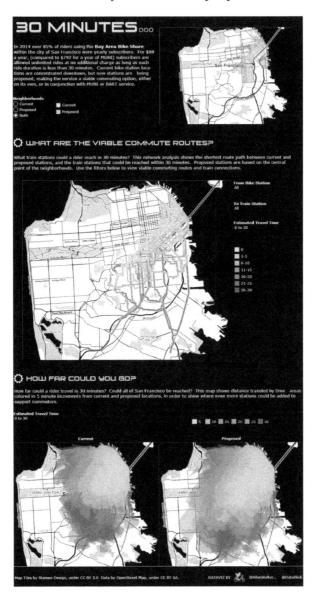

The preceding image is republished under the terms of fair use. It was created by *Anya A'hearn* and *Allan Walker*. (Source: https://public.tableausoftware.com/views/30Minutes___BayAreaBikeShare/30Minutes___?:embed=y&:loadOrderID=0&:display_count=yes.)

As humans, we relate our experiences to each other in stories, and data points are an important component of stories. They quantify phenomena and, when combined with human actions and emotions, can make them more memorable. When authors create public interest story elements with Tableau Public, readers can interact with the analysis, which creates a highly personal experience and translates into increased participation and decreased abandonment. It's not difficult to embed Tableau Public visualizations into websites and blogs. It is as easy as copying and pasting the JavaScript that Tableau Public automatically renders for you.

Using Tableau Public increases accessibility to stories too. You can view data stories on any mobile device with a web browser and then share it with friends via social media sites such as Twitter or Facebook using Tableau Public's sharing functionality. Stories can be told with text as well as popular and tried-and-true visualization types such as maps, bar charts, lists, heat maps, line charts, and scatterplots. Maps are particularly easier to build in Tableau Public than most other data visualization offerings because Tableau has integrated geocoding (down to the city and postal code) directly into the application. Tableau Public has a built-in date hierarchy that makes it easy for users to drill through time dimensions just by clicking on a button. One of Tableau Software's taglines, *Data to the People*, is a reflection not only of the ability to distribute analyses sets to thousands of people at once, but also of the enhanced abilities of nontechnical users to explore their own data easily and derive relevant insights for their own community without having to learn a slew of technical skills.

Telling your story with Tableau Public

The Tableau Software was originally imagined in the Stanford University Computer Science department, where a research project sponsored by the U.S. Department of Defense was launched to study how people can rapidly analyze data. This project merged two branches of computer science—the understanding of data relationships and computer graphics. This mash-up was discovered to be the best way for people to understand and sometimes digest complex data relationships rapidly and, in effect, help readers consume data. This project eventually moved from the Stanford campus to the corporate world, and Tableau Software was born. The usage and adoption of Tableau has since skyrocketed. At the time of writing this book, Tableau is the fastest growing software company in the world and now, Tableau competes directly with older software manufacturers for data visualization and discovery, such as Microsoft, IBM, SAS, Qlik, and Tibco, to name a few.

Most of these are compared by Gartner in its annual Magic Quadrant. For more information, visit `http://www.gartner.com/technology/home.jsp`.

Tableau Software's flagship program, Tableau Desktop, is a commercial software used by many organizations and corporations throughout the world. Tableau Public is the *free* version of Tableau's offering, and it is typically used with nonconfidential data either from the public domain or the one that we collected ourselves. This free public offering of Tableau Public is truly unique in the business intelligence and data discovery industry. There is no other software like it—powerful, free, and open to data story authors.

There are a few terms in this book that might be new to you. You, as an author, will load your data into a workbook, which will be saved into the Tableau Public Cloud.

A visualization is a single graph. It, typically present on a worksheet. One or more visualizations can be on a dashboard, which is where your users will interact with your data.

One of the wonderful features about Tableau Public is that you can load data and visualize it on your own. Traditionally, this has been an activity that was undertaken with the help of programmers at work. With Tableau Public and new blogging platforms, nonprogrammers can develop data visualization, publish to the Tableau Public website, and embed the data visualization on their own website. The basic steps to create a Tableau Public visualization are as follows:

- Gather your data sources, usually in a spreadsheet or a `.csv` file

- Prepare and format your data to make it usable in Tableau Public

- Connect to the data and start building data visualizations (charts, graphs, and other objects)

- Save and publish the data visualization to the Tableau Public website

- Embed your data visualization in your web page by using the code that Tableau Public provides

Tableau Public is used by some of the leading news organizations across the world, including *The New York Times*, *The Guardian* (UK), *National Geographic* (US), the *Washington Post* (US), the *Boston Globe* (US), *La Informacion (Spain)*, and *Época* (Brazil). In the following sections, we will discuss installing Tableau Public. Then, we will take a look at how we can find some of these visualizations out there in the wild so that we can learn from others and create our own original visualizations.

Installing Tableau Public

Now, let's look at the installation steps for Tableau Public:

1. To download Tableau Public, visit the Tableau Software website at http://public.tableau.com/s/.

2. Enter your email address and click on the **Download the App** button located in the middle of the screen, as shown in following screenshot:

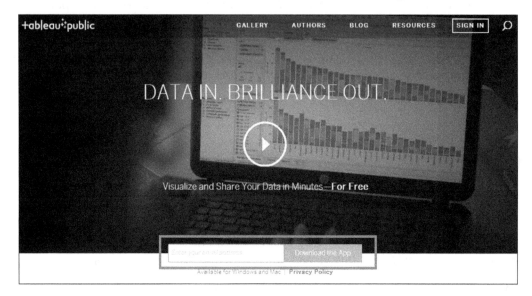

 The downloaded version of Tableau Public is free and not a limited release or demo version. It is a fully functional version of Tableau Public.

3. Once the download begins, a **Thank You** screen gives you the option of retrying the download in case it does not automatically begin or it is downloading a different version. The version of Tableau Public that downloads automatically is the 64-bit version for Windows. Users of Mac should download the appropriate version for their computers, and users with 32-bit Windows machines should download the 32-bit version.

 Check your Windows computer system type (32- or 64-bit) by navigating to **Start | Computer** and right-clicking on the **Computer** option. Select **Properties**, and view the **System** properties. 64-bit systems will be noted as such. 32-bit systems will either state that they are 32-bit systems, or not have any indication of being a 32- or 64-bit system.

4. While the Tableau Public executable file downloads, you can scroll to the lower part of the **Thank You** page to learn more about the new features in Tableau Public 9.0. The speed with which Tableau Public downloads depends on the download speed of your network, and the 109 MB file usually takes a few minutes to download.

5. The `TableauPublicDesktop-xbit.msi` (where x has a value of either 32 or 64 depending on the version that you selected) file is downloaded. Navigate to that `.msi` file in Windows Explorer or the browser window and click on **Open**. Click on **Run** in the **Open File - Security Warning** dialog box that appears in the following screenshot. The Windows installer starts the Tableau installation process:

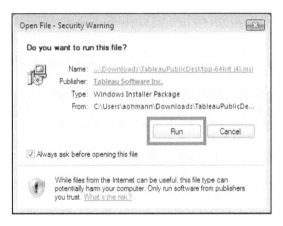

6. Once you have opted to **Run** the application, the next screen prompts you to view the License Agreement and accept its terms:

7. If you wish to read the terms of the license agreement, click on the **View License Agreement...** button.

   (You can customize the installation if you want to. Options include the directory in which the files are installed as well as the creation of a desktop icon and a Start Menu shortcut (for Windows machines). If you do not customize the installation, Tableau Public will be installed in the default directory on your computer, and the desktop icon and the Start Menu shortcut will be created.)

8. Select the checkbox named **I have read and accept the terms of this License Agreement** and click on **Install**.

9. If a **User Account Control** dialog box appears with the **Do you want to allow the following program to install software on this computer?** prompt, click on **Yes**:

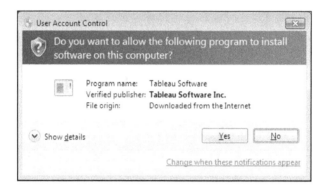

10. Tableau Public will be installed on your computer, with the status bar indicating the progress of the installation, as shown in the following screenshot:

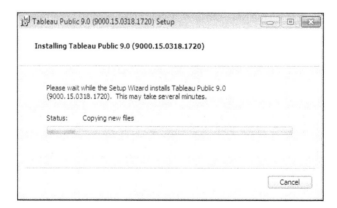

11. When Tableau Public has been installed successfully, the **Home** screen opens. The next section discusses its features.

Exploring Tableau Public

The Tableau Public home screen, as shown in the following screenshot, has several features that allow you to **Connect** to data, **Open** workbooks, and **Discover** the features of Tableau Public:

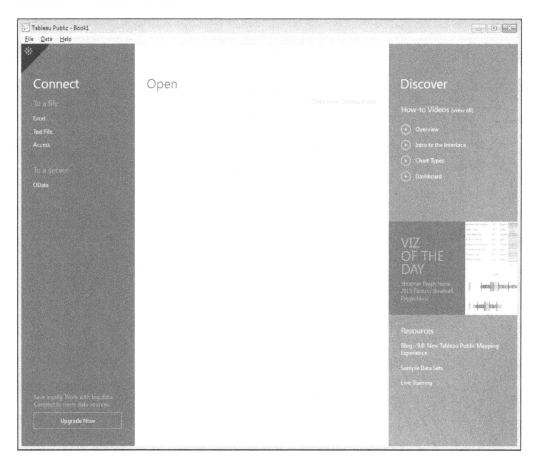

Tableau encourages new users to watch the video on this welcome page. To do so, click on the button named **Watch the Getting Started Video**. You can start building your first Tableau Public workbook any time.

Connecting to data

You can connect to four different data source types in Tableau Public, as shown in the next screenshot, by clicking on the appropriate format name:

- Microsoft **Excel** files
- **Text** Files with a variety of delimiters
- Microsoft **Access** files
- **OData** files

Chapter 2, Tableau Public Interface Features, focuses on connecting to data sources and explains this in detail.

Opening files and creating your profile

You can open the files that you create in Tableau Public by clicking on the **Open from Tableau Public** link. When you click on the link, Tableau Public will prompt you to log in with the e-mail address that you have used to create your account, as shown in the following screenshot:

When you enter your e-mail and password, Tableau Public will verify it. Then, you will be able to select the file that you would like to open.

The list of files includes the names, modification dates, and size of the workbooks that you have saved in Tableau Public. It also includes the ability for you to search by entering a string of text.

When you find the workbook that you would like to open, click on **Open**, and then the most recently saved version will open in Tableau Public on your computer, as shown in the following screenshot:

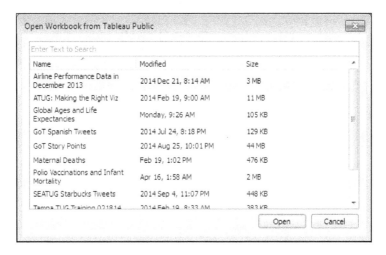

If you have not created an account, click on the link at the bottom of the screen that says **Create one now for free**.

The **Create a Profile** screen requires you to enter information in several fields, as shown in the following screenshot:

Now, let's look at each of these fields:

- **Name**: Your **Name** will be displayed in your profile. You can edit this later if you want.
- Your **Email Address** is the identifier that you will use to log in to Tableau Public.
- **Choose a Password**, which must consist of at least six characters.
- **Confirm** your password.
- **Prove You're Not a Robot**. A CAPTCHA is generated for you to verify that you're not a robot when you click on it.
- **Review the Legal** requirements and agree to the terms of service.
- Click on **Go to My Profile** to complete the creation of your profile.

When you click on to **Go to My Profile**, your web browser will open your new profile page on Tableau Public. This is a page that displays information that you enter about yourself and your interests as well as a photograph of your choosing and links to other websites with which you're affiliated.

Your profile page also displays and allows you to manage your Tableau Public workbooks. We will discuss the profile in greater detail in *Chapter 9, Publishing Your Work*.

Discover

The right pane of the Tableau Public 9.0 home screen, as shown in the following screenshot, has several features that help you learn how to use the Tableau Public 9.0:

Let's take a look at each of these features:

- **How-to Videos**: Tableau has a wealth of online videos. You can view them by clicking on the video names in the pane.

- If you would like to explore other videos, click on the **view all** link next to the header. This will open Tableau's training video section of their corporate website in your browser. If the page doesn't open, you can access it by visiting `https://public.tableau.com/s/resources`.

- **VIZ OF THE DAY**: Tableau Public's staff selects a **VIZ OF THE DAY** from the recent publications on Tableau Public. These are the visualizations that are relevant to current events, explore important questions, and/or innovatively use the functionality of Tableau Public. You can subscribe to the **VIZ OF THE DAY** and view other selections by visiting `https://public.tableau.com/s/gallery`.

Resources that you can open include the Tableau Public blog, sample Data Sets, and links to live training. You can view all of these on Tableau Public's resources page in your Internet browser by visiting `https://public.tableau.com/s/resources`.

Exploring the visualizations of other authors

We often learn by viewing other people's work. So, let's take a look at a few data visualizations created by other authors. Note that most Tableau Public data visualizations allow you to download the entire workbook. If data is not readily downloadable on the workbook page, you can export the underlying data to Excel while inside the workbook by using the desktop client of Tableau Public. There are several great places to find the best Tableau Public data visualizations, including Tableau Public and the **VIZ OF THE DAY** galleries (for more information, visit `https://public.tableau.com/s/gallery`) and the Tableau Public blog (to have a look at the blog, visit `https://public.tableau.com/s/blog`).

To make use of a recommended authors and profile finder, visit `https://public.tableau.com/s/authors`.

The Tableau Public gallery is an excellent place to look at examples of works of others, and the Tableau Public team has curated a set of popular visualizations by topic and number of views.

The recommended authors page (`https://public.tableau.com/s/authors`) is a fun place to look at both well-known and established Tableau Public authors (bloggers, journalists, and the Tableau staff) as well as lesser known authors to explore their work. You can also access an author's profile page and see their work by clicking on the **View Profile** button under their name, as shown in the following screenshot:

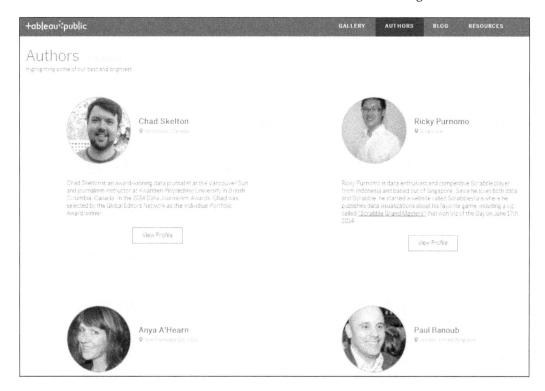

Summary

In this chapter, we had a look at how Tableau Public is commonly used. We also discussed how to download and install Tableau Public, explore Tableau Public's features and learn about Tableau Public, and find other authors' data visualizations using the Tableau Galleries and Recommended Authors/Profile Finder function on the Tableau website. In the next chapter, we will explore data connections and manipulations in Tableau Public.

2
Tableau Public Interface Features

The user interface for Tableau Public was created to be simple and intuitive. It comes with three primary features (as discussed in *Chapter 1, Getting Started with Tableau Public*), namely connecting to data, opening your work, and discovering Tableau Public. Since Tableau Public is a tool for data discovery as well as data visualization, the interface is designed to encourage discovery through the drag-and-drop features for data. The user interface for Tableau Public is segmented into separate areas, namely data elements, cards, shelves, and the canvas. The data is also divided into two general categories—dimensions and measures. By understanding how data interacts with the user interface, you can design, configure, and polish chart objects that will be built into worksheets. These worksheets can then be assembled into one or more dashboards.

In this chapter, we will cover the following topics:

- The Tableau Public user interface
- The side bar, including the **Data** window and the **Analytics** pane
- The toolbars and menus
- The **Columns**, **Rows**, and **Filters** shelves
- The **Marks** card
- The **Filters** and **Pages** shelves
- The **ShowMe** card

Touring the Tableau Public user interface

- In the previous chapter, we discussed how to download and install Tableau Public. We also saw what the start screen looks like and how you can use it to connect to data, explore your own work or that of others, or discover how to use the tool. On opening either a data file or an existing workbook with Tableau Public, you will see the worksheet view.

There can be one or more worksheets in a workbook. Tableau Public extends this further by allowing you to place one or more worksheets in a dashboard, with all of this contained within the workbook.

The starting point that appears when opening a new Tableau workbook is the worksheet view. This is the working area where you can build your dashboard. Let's take a quick look at it.

The visualization shown in the following screenshot uses data from the **Federal Aviation Administration** (**FAA**) on every commercial flight at the domestic airports of the United States in March 2015 to average the departure delay (in minutes) for every weekday. You can download this data from http://www.transtats.bts. gov/DL_SelectFields.asp?Table_ID=236&DB_Short_Name=On-Time. Also, you can download the companion Tableau Public workbook from the profile of this book at https://public.tableau.com/profile/tableau.data.stories#!/. We will use this workbook to explore parts of the Tableau Public interface, as shown in the following screenshot:

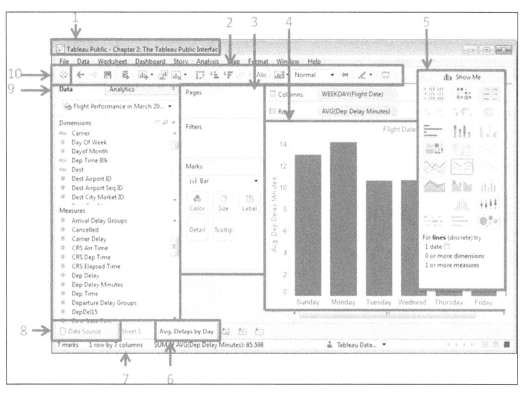

The following are the parts of the user interface that are shown in the preceding screenshot:

- **Workbook (1)**: This is the workbook title, which is the name given to the workbook when you save it

- **Toolbar (2)**: This is where you can save your work, among other functions

- **Cards** and **shelves (3)**: These are the areas where you can add fields or filters to the visualization

- The **View**, or the **Visualization (4)**: This is the graph itself

- The **ShowMe** card **(5)**: This prompts you to create different visualization types based on the data selected

- **Sheet tabs (6)**: This allows you to create, rename, or duplicate sheets and dashboards

- The **Status** bar **(7)**: This shows the aggregated totals of the marks on your visualization

- **Data Source (8)**: Links back to data sources
- The **Sidebar (9)**: This contains both the Data window and the Analytics pane
- The **Start** button **(10)**: This takes you back to the home screen

The side bar

Our discussion about the parts of the user interface starts with the side bar because it contains both the **Data** window and the **Analytics** pane. First, we will talk about the **Data** pane. After that, we will discuss the **Analytics** pane.

The Data pane

The Data pane is where your data sources load. In addition to listing the fields in your data source either alphabetically or by folder, the **Data** pane includes visual cues that tell you the type of each field. The following screenshot shows the visual cues of the **Data** pane:

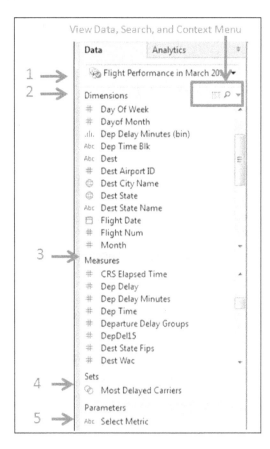

Tableau Public scans the contents of your data source, groups fields into **Dimensions**, and **Measures** according to their field type. Before you start working with your data, you should look through the fields in the **Dimensions** and **Measures** panes to make sure that each field is in the proper place.

The **Data** pane has the following five different sections:

1. The **data source name (1)**: When you load data, you should provide the data source with a name that identifies the contents, because that is what you and your consumers will see when they download your workbook from Tableau Public. Once you have added several data sources, you can condense their window in order to save space and then select different sources from the drop-down menu.

2. The **Dimensions** pane **(2)**: This includes categorical fields with qualitative data. The **Dimensions** pane typically consists of a string field, a date field, and a field that has geographical attributes, as well as unique identifiers, such as ID fields.

3. The **Measures** pane **(3)**: This usually includes quantitative fields with numerical data that can be aggregated. Tableau Public will automatically group numerical fields, except the ones with the ID string in the name as measures.

4. The **Sets** pane **(4)**: This includes user-defined, custom fields that interact just like dimensions and measures do. **Sets** pane can also create subsets of data that you can use just like dimensions.

5. The **Parameters** pane **(5)**: This includes dynamic placeholders that can replace constant values in calculated fields and filters. Parameters are unique to a workbook and not a data source. You'll see the parameters available in your workbook no matter which data source you are viewing.

From the **Data** pane, you can can create fields for the **Data** window, as follows:

1. Right-click on the data source to edit it.

2. Click on the **View Data** icon to see a sample of your data set.

3. Click on the search icon (the little magnifying glass) to search for fields.

4. Click on the arrow that points downwards, which is the **Context** menu, to create calculated fields, parameters, and change the sort/view options for the **Data** window.

Visual cues

Within the **Data** pane, each field name is displayed, but there is also a visual cue that tells you what type of field it is. Field types determine the function and capability of joining data sources. In addition to showing the field type, Tableau Public allows you to change a field's type by right-clicking on it and selecting the **Change Data Type** option. The following list shows the fields and their descriptions:

- The **Abc** field: This indicates that the field is a string field, which means that the contents of the field may include letters, special characters, or numbers

- The calendar icon: This indicates a date, datetime, or time field

- The # sign: This indicates a numeric field that can have any type of native numeric format, from `bigint` to `decimal`

- The globe icon: This icon indicates that the field has geographic attributes

- The paper clip icon: This indicates that the field is a group that you have created in Tableau Public

- The Venn diagram icon: This indicates a set

 The parameters have their own data types, which can be established when you create them.

The Analytics pane

The **Analytics** pane is next to the **Data** pane. You can access it by clicking on its header. It provides you with the ability to add summaries, average lines, constant lines, distribution bands, medians, boxplots, forecasts, and other visual analytics to your visualization. You can then customize and edit them by using the reference line and formatting user interfaces.

The **Analytics** that you can add are dependent on the data elements in your visualization, which include the following:

- **Summarized Analytics**: This includes the following elements:
  - A constant line: This is an integer on an axis that you can input
  - An average line: This displays the mean of the measure that you have selected either across the table, pane, or cell
  - Median line with 25% and 75% quartiles: This creates a median reference line and a quartile distribution band that you can edit
  - A box plot
  - Totals and subtotals

- **Modeling**: This includes the following elements:
  - ○ **Average** (or mean) with confidence intervals
  - ○ Trend lines with the most commonly used models (such as linear, exponential, and so on)
  - ○ Forecast
- **Custom**: With the help of this element, you can add custom reference lines and distribution bands as well as box plots, which can also be used by right-clicking on an axis

In the following screenshot, we added the average delay in minutes per day to the graph as a reference line by performing the following operations:

1. Click on **Analytics** to see the **Analytics** pane.
2. Drag **Average Line** to the *y* axis, which is the vertical axis.
3. Select **Table** as the scope, as shown in the following screenshot:

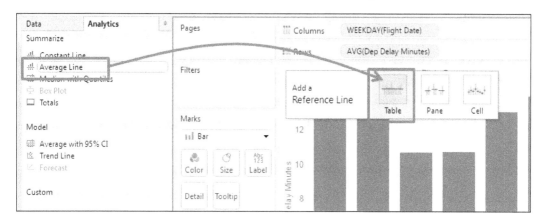

Menus and toolbars

The menus in Tableau Public are arranged and named in a way that is similar to those in other modern applications. Their primary uses are shown in the following screenshot:

Let's look at each of them in the following list:

- **File**: You can open and save your work to Tableau Public via this menu. Remember that Tableau Public does not auto-save.
- **Data**: From here, you can add new data sources and modify the existing ones.
- **Worksheet**: From here, you can create new worksheets, copy the visualization or data on your worksheet, and modify the title, caption, tooltip, and other features.
- **Dashboard**: This is used to create and format dashboards as well as to add actions.
- **Story**: This is used to add **Story Points**, which enhance the narrative capabilities of a data visualization. **Story Points** refers to a specific function in Tableau Public, and it's beyond the scope of this book.
- **Analysis**: This can be used to aggregate and disaggregate measures, create forecasts, totals, and trend lines, and create and edit calculated fields.
- **Map**: By changing the **Map** options, you can modify background maps and images, and add custom geocoding and WMS.
- **Format**: This is used to modify the appearance of visualizations.
- **Window**: This can be used to switch between the presentation and development mode as well as other views in the workbook
- **Help**: You can use this to get help and manage performance.

The buttons on the toolbar are graphically descriptive of their function. When you roll over each with your mouse, the instructions for use pop up as well.

The buttons of the toolbar include the following:

- The Start button: This takes you back to the Start screen
- The Undo button: This reverses the previous action that you took and can go as far as back to the state of the workbook when you opened it
- The Redo button: This repeats actions that you have reversed
- The Save button: This is critical, as Tableau Public does not auto-save
- Add New Data Source: This adds a new data source
- Add New Worksheet: This adds a new worksheet
- The Duplicate button: This duplicates the current worksheet
- The Clear button: This clears the current worksheet
- The Swap button: This swaps the fields on the **Rows** and **Columns** shelves

- Ascending: Sorts in ascending order
- Descending: Sorts in descending order
- Group: This is inactive in the view, as shown in the previous screenshot
- Show Mark Labels
- The Reset cards: These allows you to show legends and cards that you might have hidden or removed
- Fit: This allows you to change the fit of the visualization within a window
- Fit Axes: This allows you to set axis ranges
- Highlight: This allows users to click on dimension members and highlight the related records in other visualizations on the dashboard
- The Presentation mode: This hides the menus, the **Data** pane, or the **Dashboard** pane

Canvas and Column/Row shelves

Tableau Software has a user interface that is very different from that of the older reporting or data analysis tools that you may have used at work or in school. It uses a methodology of dragging and dropping objects for most of the functions that you need to perform to build a visualization. The areas of the workspace where you place objects are called **Shelves** and **Cards**. Many of the tasks in this book instruct you to drag a field to the **Columns** or **Rows** shelf.

The far right side of the screen is called the canvas area. It is where sheet objects, such as a chart, are built. The chart area itself starts out blank, and you must drag the fields that you want to analyze to an axis, header, or the **Columns** or **Rows** shelf, which also determines on which axis or header the field appears.

In the following screenshot, we have highlighted the **Columns** and **Rows** shelves, the canvas, the **Marks** card, the **Filters** shelf, and the **Pages** shelf. This screenshot shows how the workspace looks before we add fields to it, and the description is as follows:

- The **Columns** Shelf **(1)** is where you put fields that you want on the horizontal header or the x axis, that is, the axis that goes from left to right. The **Rows** shelf is where you put fields that you want on the y axis.

- The **Canvas (2)** has three places where you can drop fields, namely the x axis/header, the y axis/header, and the visualization space itself.

- The **Marks** card **(3)** includes the controllers that are used to mark color, size, text label, and shape. In addition to this, the level of the **Detail** and **Tooltip** controllers allows you to control the appearance of the visualization. You can change the mark type for each individual axis as well as control the size, color, label, and shape. (Since the **Marks** card is critical to the function of the visualization, we will describe it in depth in the next section).

- The **Filters** shelf **(4)** is where you put fields that you want to use to limit the values included in your visualization. After dragging a field onto the **Filters** shelf, you can select values in the dialog box of that data element's Filters shelf.

- The **Pages** shelf (**5**) allows users to progress through your visualization based on the fields that you put on it. For instance, you can use a date field on the **Pages** shelf, and your users can click on changes over time without having to manually select the subsequent values.

Using the Columns and Rows shelves

As discussed earlier, the fields that you place on the **Columns** and **Rows** shelves will appear either on the x and y axis, or the row or column headers of the visualization respectively. The following are a few new concepts related to the uses of these shelves:

- Once a field is on the shelf, it is referred to as a **pill**, or an **active field**
- A pill on a shelf has a context menu

 The context menu can be opened by clicking on the small arrow that point downwards.

- Instances of fields that are discrete will appear in blue, and continuous fields will appear in green. The following is a description of both these fields:
  - ○ Discrete fields have specific values, and the range of values is finite
  - ○ Continuous fields, on the other hand, have an infinite range of values with infinite possibilities
- The Information Lab blog does a great job explaining the differences in Tableau Public; you can have a look at the explanation by visiting `http://www.theinformationlab.co.uk/2011/09/23/blue-things-and-green-things/`
- In case you have multiple pills on a shelf, the discrete pills appear first because they are used to group fields, and the continuous pills appear second because they are used to measure fields

The following screenshot shows an example of a basic graph that we created:

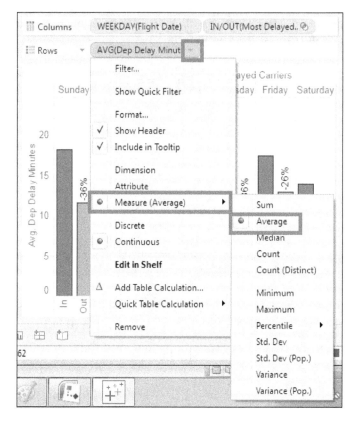

In order to create the graph shown in the preceding screenshot, we will perform the following steps:

1. Drag **Flight Date** to the **Columns** shelf from the **Dimensions** pane.
2. Choose **WEEKDAY** as the part of the data that we want to show.
3. Add a set that we created that groups airlines by the criteria that we established.

4. When we drag the **Flight Date** field and the set to the **Columns** shelf, they both appear in the column headers. Drag the **Delay** minutes from the **Measures** pane to the **Rows** shelf.

5. When you do this, it defaults to aggregating as a **SUM**. Click on the **Context** menu of the pill and change the aggregation to **Average**.

Next, we will discuss how to use the **Marks** card to add color and labels.

Using the Marks card

Another Tableau invention is the use of cards. Cards are containers for various controllers: which are dialogs in the Tableau workspace that allow various data elements and components to be configured. The most important card is the **Marks** card, which is in the most current version of Tableau Public. It has combined various controllers into one.

The **Marks** card is a compact yet highly functional area of the worksheet view that contains different controllers for data element chart properties (these data points on a chart are called marks). To use the **Marks** card, drag and drop data elements onto a corresponding shelf (such as **Colors**, **Label**, and **Size**). This will change the chart visualization by changing the chart mark properties.

The different controllers on the **Marks** card, which are commonly referred to as shelves, include the following:

- **Colors**: This changes the colors of marks in the chart (such as bar or line colors). In our example, the **Set** that shows whether an airline is in the top five worst airlines according to their average delay time is on the Color shelf.

- **Size**: This configures the sizes of data points or elements in charts.

- **Label/Text**: This adds a label to the chart for data points, bars, groupings, or lines. In our example, the percentage difference in average delay time between the top five worst airlines and all the others is on the **Label** shelf.

- **Details**: This adds details to the chart or data points, allowing you to keep the main structure of the chart. However, it further categorizes it in detail and with more granularity. For example, if we add **Unique Carrier** to the **Detail** shelf, our visualization will aggregate the average departure time by day for each **Unique** carrier, which means that there is now a bar segment for each carrier, and the story that our graph tells will change dramatically, as shown in the following screenshot:

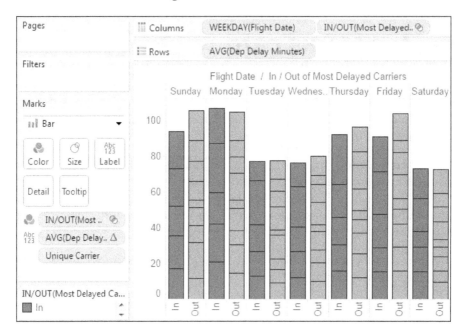

- **Tool Tip**: This allows you to add context and calls to action, which are critical for telling stories because they instruct users how to progress to the next step and they better illustrate how a data point relates to a user's interests.
- **Shape**: This sets the data point shape in the chart visualization.
- **Path**: This is typically used for routes on a map. This controller allows a path to be sequentially built and placed on a map visualization. This is commonly seen in tornado and hurricane tracking maps visualizations.

Once a field is on a shelf on the **Marks** card, it shows up under the shelves with a small icon for the shelf type followed by the field name, as shown in the following screenshot:

You can change the shelf on which a field is placed by clicking on the small icon and then selecting the shelf to which you'd like to move the field, as shown in the previous screenshot.

The Filters and Pages shelves

The **Filters** shelf is where you drag fields to limit the data points that your users see in your visualization. When you drag it to the **Filters** shelf, Tableau Public will prompt you to filter a field according to its type.

You can also filter a visualization based on the inclusion in a **Data** set. If you want to show the filters to users, you can right-click on the fields in the data window to **Show Quick Filters** as well. We will describe filtering in depth in future chapters.

In the following screenshot, we have added **WEEK(Flight Date)** to the **Pages** shelf. The controller appears here. We can set the speed at which the visualization changes. In the Tableau Public desktop client, the visualization will progress sequentially once you click on the **Play** button. However, online, it will not progress automatically. Your user will need to click on the **Play** button to proceed.

Since the functionality is limited online and that is where your users will interact with your work, we will condense the discussion of the **Page** shelf:

The workspace control tabs of Tableau Public

The tabs at the bottom of the Tableau Public screen are either sheet or dashboard tabs. The tabs with a series of rows and a column on them are **sheet tabs**, where you can design objects that fit on one sheet. The tabs with a rectangle divided into four quadrants are **dashboard tabs**, where you can add one or more sheets in order to form a unified dashboard. When you click on one of the new (blank) worksheet or dashboard tabs, that tab opens and names the sheet or dashboard the next sequential name, such as **Sheet 3** or **Dashboard 2**. The sheet or dashboard is blank when you first open a new tab.

A dashboard is really just a group of worksheets on the same page. You can assemble a dashboard after creating the various component worksheets that it will contain. Dashboards are assembled and configured by using the **Dashboard** tabs in Tableau Public, which are available in the lower row of the tabs in the Tableau Public interface.

The lower-right hand side of the Tableau Public interface also contains three workspace control tabs, as shown in the following screenshot:

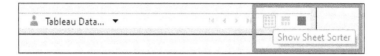

Let's take a look at each of these tabs:

- **Show Sheet Sorter**: This shows the sheets in your workbook and allows you to reorder them easily
- **Show Filmstrip**: This shows the visualization with the thumbnails of other worksheets, which is shown in the previous screenshot
- **Show Tabs**: This shows the default view of the visualization and the named tabs of other worksheets, as shown in the previous screenshot

The Show Me tool

A very useful tool in Tableau Public is the **Show Me** tool, which is available as a floating window when you click on the **Show Me** button on the upper-right hand side of the Tableau Public interface.

This tool, as shown in the following screenshot, is a helpful aid when you wish to select various chart types that can be used with your data:

 The **Show Me** tool is an optional tool. Charts can be created without any help from this tool.

The **Show Me** tool also contains a list of tips to add data to shelves and cards, such as the number of dimensions and measures necessary to create the desired chart type.

The **Show Me** tool also suggests graph types based on the view that the user has created and/or the fields that the user has selected from the **Data** window. Based on the selections and the chart type selected in **Show Me**, the tool also instructs the user what field types are needed to create a certain graph type.

For example, selecting a continuous date field will make the **Show Me** tool prompt the suggestion of a line graph. Selecting a geocoded or latitude/longitude field will make it prompt the selection of a geographic map.

In *Chapter 4, Visualization - Tips and Types*, we will discuss the **Show Me** tool in detail and have an overview of the chart types that are available in Tableau Public. Many of these chart types are covered by the **Show Me** tool suggestions.

 The **Show Me** tool dialog does not auto-hide once the user makes a selection from it. Click on the **Show Me** button again to hide the **Show Me** tool dialog.

Summary

In this chapter, we learned the Tableau Public user interface, from the welcome screen to the worksheet and dashboard tab views. We discussed the concepts of shelves and cards, walked through an example of how the Marks card affects a data visualization, and had a look at how to create a dashboard from the various worksheets created in a Tableau Public workbook. Lastly, we discussed the Show Me tool and how it can aid you in choosing appropriate data visualization types for your data.

In the next chapter, we will discuss the various chart types that you can create in Tableau Public and what some of the best practices and uses are for the chart types.

3
Connecting to Data

Visualizations depend on the data in them, and however aesthetically pleasing your visualization might be, it may be misleading or even wrong, unless the data has been formatted, aggregated, and properly represented.

This chapter discusses the major elements of finding, cleansing, understanding, formatting, and aggregating data that you will need to understand in order to produce accurate visualizations that tell compelling stories, including the following elements:

- Where you can get publicly available data and how to use it
- What tables and databases are
- The data formats that Tableau Public connects to
- Databases, tables, dimensions, facts, and field formats and conventions
- Preparing data to load it into Tableau
- Connecting to the data from Tableau Public
- Using the data interpreter
- Pivoting fields

Public data

The data sets that are publicly available or the ones that you have compiled on your own, are ideal for Tableau Public. Since all users will be able to download this data and create their own visualizations once you have published your workbook, your data set should not contain information that may be considered sensitive, which can be anything that can be used to identify a private individual or reveal confidential corporate information or intellectual property.

Public data is readily available online. Tableau Public maintains a catalog of publicly available data. Much of this data is produced by various governments, economic groups, and sports fans, along with a link to, and a rating for each source. This catalog is updated monthly, and it is a great introduction to using publicly available data. You can find it at http://public.tableau.com/s/resources.

The **Google Public Data Explorer** has a large collection of public data, including economic forecasts and global public health data. This tool is unique because it allows users to make simple visualizations from all the original data sources without having to investigate the source data, though most of it is available by linking available resources.

There are several tools available for the scraping of data from public sites too, such as ScraperWiki, import.io, and IFTTT, among others. These industries and such tools pertaining to the industry evolve rapidly. Therefore, we will not discuss any specific tool. Social media applications, such as Twitter, have made it possible for individuals and companies to build application programming interfaces (APIs) to connect to their data streams. This is useful for nonprogrammers because it's often free of cost if you wish to scrape data about specific topics, hashtags, or users with a minimal amount of coding.

Not all data is public data, and it's very important to determine whether a data set is public before using it. None of us wants to end up being sued, with a ruined reputation, or as a victim. If your source data set has identifying characteristics, first and last name, address, and financial, geolocation, medical, or federally or state protected data, it should be removed or de-identified and then saved separately before saving the visualization to Tableau Public (or not used at all). Each state has guidelines on what is considered protected information, and it's a good idea to check the restrictions in case there's even the slightest chance that your data set has sensitive information in it.

Additionally, data from a corporation should never be used unless the corporation has given the permission to use it. (Did we mention lawsuits? Being fired also isn't any fun.)

Tables and databases

Once you have found the data set that's ideal for your visualization, it's helpful to know how data stores are structured and what the different terms are.

Data is stored in tables. A table is an array of items, and it can be as simple as a single word, letter, or number, or as complicated as millions (or more) of rows of transactions with timestamps, qualitative attributes (such as size or color), and numeric facts, such as the quantity of the purchased goods.

Both a single text file of data and a worksheet in an excel workbook are tables, though this may not be apparent. When grouped together in a method that has been designed to enable a user to retrieve data from them, they constitute a database. Typically, when we think of databases, we think of the **Database Management Systems** (**DBMS**) and languages that we use to make sense of the data in tables, such as Oracle, Teradata, or Microsoft's SQL Server. Currently, the Hadoop and NoSQL platforms are very popular because they are comparatively low-cost and can store very large sets of data, but Tableau Public does not enable a connection to to these platforms. They are considered enterprise tools that should be used with Tableau Desktop Professional. Therefore, our discussion about these tools is limited.

Tableau Public is designed in such a way that it allows users in a single data connection to join tables of data, which may or may not have been previously related to each other, as long as they are in the same format. In other words, multiple CSV files or worksheets can be joined in the same excel workbook. Then, users can specify the conditions under which they need to retrieve data from the tables and how to aggregate it (examples are given in following section). Thus, that data connection becomes a de-facto database.

The most common format of publicly available data is in a text file or a **Character-Separated Values** (**CSV**) file. CSV files are useful because they are simple. The rows of data, which may or may not contain a header row, are separated by line breaks. The fields within each row can be separated by a character. Typically, this character is a comma, pipe, or tab. Commas present difficulties because the content of the fields can contain them, which causes the text to shift into a new column.

Many public data sources do allow data to be downloaded as Excel documents. The World Bank has a comprehensive collection, and we will demonstrate the connective capabilities of Tableau Public using one of its data products. Tables can be joined in Tableau Public by manually identifying the common field among the tables.

The data sources that Tableau Public connects to

Tableau Public connects to four different data sources, namely Access, Excel, text file (CSV or TXT), and OData; the first two data sources are bundled with Microsoft Office (in most cases), and the second two are freely available to everyone, regardless of the operating system that they are using. Text files are the default source of origin for most of the data that we will discuss, and anyone can create and distribute them.

Tableau Public does not connect to enterprise tools such as Teradata, Oracle, or Hadoop, and it does not connect to SQL Server Management Studio, though it does connect to flat files' output from these tools and other **Online Analytical Processing (OLAP)** systems. SQL Server Management Studio is free for noncommercial use, and it's common to use it to design basic star schemas and clean noncommercial data. Tableau Public is free because it's assumed that people use it only for their personal endeavors and not projects that generate revenue for their employers. If this is not the case, they should upgrade to the Tableau Desktop Professional edition.

The databases, tables, dimensions, facts, field formats and conventions

Data that is retrieved from different sources will invariably have different structures. Some of these data resources need more formatting than others in order to turn them into clean, usable tables.

As previously mentioned, a table might be as simple as having a single digit in a text file. As long as users know what that digit represents, they can assign a qualitative or quantitative value to it. Imagine a situation where you are collecting rainfall measurements. Entering the amount of rainfall as subsequent rows of text into a new file constitutes a table.

The amount of rainfall is a measure; it is a quantitative fact. A dimension is a field that contains qualitative data. In this case, both the time of the day and the location of the measurement will be dimensions. Dimensions are typically formatted as date, string, or character fields, while measures are formatted as numbers. Text files do not have field formats, which are considered metadata, but Microsoft Excel and Microsoft Access do contain this information.

It's important to make sure that the formats for fields of the same type (the date or the primary/foreign key) are consistent between worksheets in a workbook or tables in an Microsoft Access database, because Tableau Public automatically joins only the fields with the same format and the exact same name (including capitalization). If your field names are not the same but they should be joined using join conditions, you can join them manually.

Another common dimension is a unique identifier, which assigns a non repetitive value to each object in a set. A phone number is a unique identifier as it is related to only one phone at a time. The same is the case with a social security number. Within a data set, a person's name will be considered a unique identifier if it is not repeated; if it were, then a different unique identifier would be used to identify individual people. Thus, it's common to use numerical fields as primary keys for individuals, and these numerical fields can be used across multiple tables and across different dimensions.

Tables need to be structured so that the field names (dimensions and facts) go across and the rows of data (the dimension values and measure facts), go down the table. There are some databases that transpose data because their querying engines are optimized to search across columns rather than down the rows, but most DMBSes are not columnar, and Tableau is not built to search rows.

The following table is a great example of a table that is structured properly. This is the 2012 NFL performance data that is freely available:

Player	Team	Receptions	Yards	Average
Calvin Johnson	Det	122	1964	16.1
Wes Welker	NE	118	1354	11.5
Brandon Marshall	Chi	118	1508	12.8
Andre Johnson	Hou	112	1598	14.3
Jason Witten	Dal	110	1039	9.4
Reggie Wayne	Ind	106	1355	12.8
A.J. Green	Cin	97	1350	13.9
Demaryius Thomas	Den	94	1434	15.3
Tony Gonzalez	Atl	93	930	10
Sharod White	Atl	92	1351	14.7

Each column is a field; the dimensions are **Player** and **Team**, and the measures are **Receptions**, **Yards**, and **Average**. The primary key is `Player`. There is only one row for each player. `Team` is the foreign key, as it may be the primary key in other tables, such as the aggregations by team.

Conversely, knowing what not to do is as instructive as knowing what to do. The following table, which shows the population (in millions) by country, is a good example of what not to do:

Country	1950	1951	1952	1953
Afghanistan	8,151	8,277	8,407	8,543
Africa	229,895	234,594	239,501	244,621
Albania	1,215	1,240	1,269	1,303
Algeria	8,753	8,953	9,141	9,326
American Samoa	19	19	20	20
Andorra	6	7	7	8
Angola	4,148	4,219	4,297	4,377
Anguilla	5	5	5	6
Antigua and Barbuda	46	48	50	51
Argentina	17,150	17,506	17,865	18,224

The problem with this table is that the years, which actually are qualitative descriptions of when each population measurement was made, are used as separate columns even though the year is a dimension and should run down the page. (Dates are dimensions too and not facts, because they describe something qualitative). If we loaded this table in Tableau Public, we would see a separate measure field for each year because Tableau Public recognizes each column as a distinct field. (In one of the following examples, we will use Tableau Public's new data interpreter to structure the data source properly).

The correct structure for this table will have three columns, namely [Country], [Year], and [Population], with a separate row for each combination of country and year.

Connecting to the data in Tableau Public

Tableau Public has a graphical user interface (GUI) that was designed to enable users to load data sources without having to write code. Since the only place to save Tableau Public documents is in Tableau's Cloud, data sources are automatically extracted and packaged with the workbook. (The ability to save extracts as separate documents or open extracts and share them with different users is a feature of Tableau Desktop Professional).

Connecting to data from a local file, that is, an access, excel, or text file saved on your computer, takes several steps that have little variability by data source, which will be illustrated as follows with detailed screenshots:

1. Click on the **Connect to Data Link** option from the **Data** menu.

2. Select the data source type.

3. Select the file or website to which you want to connect.

4. For a Microsoft Access, Microsoft Excel, or a text file, determine whether the connection is to one table or multiple tables or it requires a custom SQL connection:

   ° If the connection is to one table, select the table.

   ° If the connection is to multiple tables, select the option for the connection to multiple tables and identify the join conditions. We will discuss this in detail in the next section.

   ° Alternatively, you can type or paste a custom SQL.

5. When all the selections have been made, click on **Ok**.

Now that you have learned what data sources look like and how they are structured, we will give you a couple of examples of data connections.

In the first exercise, we will connect to the World Bank's environment indicators. You can download this data, which is formatted either for Microsoft Excel or as a text file, at www.worldbank.org.

This data source is not formatted properly, which is shown in the following screenshot. It has a spacer row between the top of the worksheet and the headers' rows, and the years in which the measurements were taken were distributed as columns rather than individual dimensional values in a row.

Once we connect to the file, we will use Tableau Public's new data interpreter to clean up and pivot the rows:

	A	B	C	D	AS	AT	AU
1	Data Source	World Development Indicators					
2							
3	Country Name	Country Code	Indicator Name	Indicator Code	2000	2001	2002
4	Aruba	ABW	Land area where elevation is below 5 meters (% of total lan	AG.LND.EL5M.ZS	29.57481		
5	Aruba	ABW	Forest area (sq. km)	AG.LND.FRST.K2	4	4	4
6	Aruba	ABW	Forest area (% of land area)	AG.LND.FRST.ZS	2.222222	2.222222	2.222222
7	Aruba	ABW	Cereal production (metric tons)	AG.PRD.CREL.MT			
8	Aruba	ABW	Access to electricity (% of population)	EG.ELC.ACCS.ZS	84.99329		
9	Aruba	ABW	Electricity production from oil, gas and coal sources (% of to	EG.ELC.FOSL.ZS			
10	Aruba	ABW	Electricity production from renewable sources (kWh)	EG.ELC.RNEW.KH			
11	Aruba	ABW	Electricity production from renewable sources, excluding h	EG.ELC.RNWX.KH			

The data source user interface

Before loading data, it's important to know what the different parts of the data connection user interface are. The Tableau Public 9.x user interface that is used for the connection to data is shown in the following screenshot. Don't forget that the Start button remains in the upper-left corner of the UI. You can click on it from either the data source window or a worksheet to get back to the Start menu:

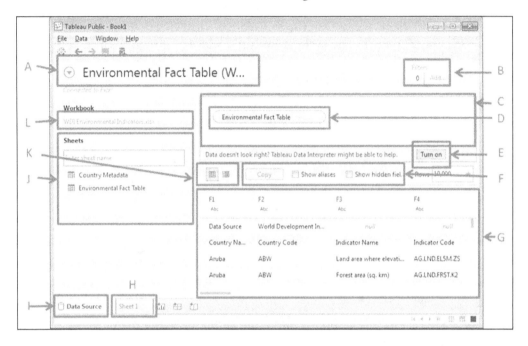

The parts of the user interface and their descriptions are as follows. The parts have been given in alphabetical references:

References	Description
A	This is the data source name, which will be modified in subsequent exercises
B	These are the data source filters, which can be used to limit the data that you load
C	This is the workspace, where you can add and join tables
D	These are individual tables
E	This is the Data Interpreter, which is available for Microsoft Excel files; we will learn how to turn it on and use it in subsequent exercises

References	Description
F	Edit data source display by showing/hiding fields
G	This is the data
H	This is a link to sheets; you can click on this to go back to your worksheets
I	This is the Data Source button, which can be clicked on from any worksheet to get back to the data source
J	These are the tables within the data source, which can be dragged to the workspace to join to other workspace
K	This is the pivot or view grid of the data, which will be used in subsequent exercises
L	This is the data source, which can be changed by clicking on the orange link and then browsing to a new file

To load this file into Tableau Public, we will start with a new Tableau Public workbook. You can download the Tableau Public workbook that we used for this chapter by visiting `https://public.tableau.com/profile/tableau.data.stories#!/`. The following steps will guide you through how to connect a file to Tableau Public:

1. Open a new instance of Tableau Public.

2. From the **Connect** pane, click on the data file type to which you'd like to connect. In this case, we are using an excel file.

3. Browse to the file to which you would like to connect.

4. Drag a table from the list of tables, which is a list of different worksheets in this case, along with the workbook onto the workspace.

5. Note that the values in the data source are now populating the space below the workspace, but at least with this data set, there is no complete set of field headers. We will edit the data source by using the data interpreter in the next exercise.

The name of the data source is showing a concatenation of the name of the workbook and the table name. Click on it (in the previous screenshot, it's **A**) to give the data source a good name. Remember that anything that you publish on Tableau Public is available for other people to download, and since they aren't able to see the actual origin of your data source, it's a good idea to give it an explicit name so that there are no errors of attribution.

Using the data interpreter

Tableau Public 9.x has a new feature that is designed to reduce the amount of transformation that you need to do to your data sources. The data interpreter automatically detects where the first row of headers or data is in a Microsoft Excel data file, and if there are empty or semi-structured rows before the data, it can remove them. (The data interpreter does not work with text files.)

The following data source has the following major errors in it:

- There are two rows of mostly blank, non-data values before the first row of the valid data; we will use the data interpreter to fix this

- The years in which the measurements were taken should be going down a column rather than across the columns; we will pivot the data in the next exercise to fix this

We will use the data interpreter to fix the first problem. In the following screenshot, you will see that Tableau Public has suggested that we use the data interpreter. The steps are as follows:

1. Under the workspace, note that Tableau Public has recognized that the data might not be formatted properly and has suggested using the data interpreter. Click on the button that says **Turn on**:

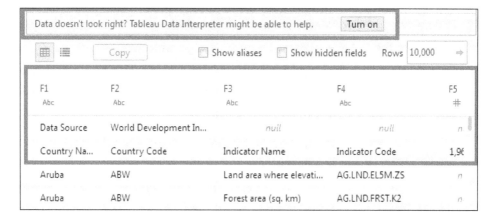

2. Now that the data interpreter is on, you can review the results in the following screenshot and see how the data was transformed:

The two rows of garbage are now gone, and the field headers are populating properly. We still need to resolve the issue of the date dimension going across the columns rather than down a column. We can resolve this by pivoting the data.

Pivoting data

Pivoting data is a capability designed to help you resolve issues within data sources, like in the previous example, where the date dimension is not formatted properly.

By highlighting the headers, you can pivot them from columns into rows by performing the following steps:

1. Highlight the field headers that you need to pivot. In this case, we click on **1960** and scroll all the way to the right, holding down the Shift key as we select columns.

2. Right-click on a selected header and choose **Pivot**.

3. The pivoted fields now have transformed into two new columns — the headers that you selected appear as values in a new column called `pivot field names`, and the measures now appear in a field called `pivot field values`.

4. Right-click on the headers for each of these fields and rename them. We renamed `Pivot field names` to `Year` and `Pivot field values` to `Measure`.

5. Check out the following modified data source. It is now formatted properly, but there is one issue—the numerous rows with null values. We will edit those in the next exercise:

Country Name	Country Code	Indicator Name	Indicator Code	Year	Measure
Environmental Fa...	Environmental Fa...	Abc	Abc Environmental Fac...	Abc	# Pivot
		Environmental Fact Table		Pivot	Pivot
Aruba	ABW	Land area where elevation is below 5 met...	AG.LND.EL5M.ZS	1960	*null*
Aruba	ABW	Forest area (sq. km)	AG.LND.FRST.K2	1960	*null*
Aruba	ABW	Forest area (% of land area)	AG.LND.FRST.ZS	1960	*null*
Aruba	ABW	Cereal production (metric tons)	AG.PRD.CREL.MT	1960	*null*

Filtering data sources

It's reasonable to expect that you won't need to load all the data in the data source. It is important to load only what you need because the more the data in the data source, the slower it will be. In the current example that we are using, there are many rows with null values. The reason that they have null values is that for the selected measure, no measurement was taken for certain time periods.

> The null values are different from measurements of zero. Values of zero mean that a measurement was taken and the value was zero. Null means that no measurement was taken.

We have no reason to load rows with null values. Therefore, we can filter them as follows:

1. In the upper-right corner under **Filters**, click on **Add**.
2. Click on **Add** again.
3. Select the field that you wish to filter. (We filtered on **Measure**).
4. Since **Measure** is a measure and not a dimension, we see a continuous spectrum of values. But we want to include everything *except* null values.
5. Click on the **Special** button on the upper-right side.
6. Click on **Non-null values**.
7. Click on **OK**.
8. Click on **OK** again.

> The data source shows values of zero, but not null. The data source is almost complete. The only item that is remaining before we can start using filter is joining it with another table in the same data source.

Joining tables

In this exercise, we will join the **fact** table with a dimension of the countries so that we can group the countries by region. A join is a logic statement in which you tell Tableau's data engine how two tables are related to each other. There are two parts to it, the join types (the left join, inner join, right join, or outer join) and the join conditions.

- **The left join**: This keeps all the records from the left (or first) table and the corresponding records from the right (or second) table.

- **The inner join**: This keeps only the records from both the tables that match the join condition.

- **The right join**: It is the opposite of the left join; it keeps all the records from the right table and only the corresponding records from the left table. Outer joins keep all the fields from all the tables.

 The availability of join types depends on your data source. For this data source, we can create an inner join or a left join.

A join condition is where you tell Tableau Public's data engine, which is functioning as a database management system in this case, how the two tables are related. In order to join tables, you need to have at least one field whose contents occur in both tables. In the following example, we will join our tables by the country name so that we can see the corresponding region for each country.

 Tableau Public will automatically join your tables on the first fields that occur alphabetically in both the tables and have the exact same field name and field type.

In order to add new tables, you need to drag them from the list of tables on the left into the workspace next to the existing tables:

1. Drag the **Country Metadata** table into the workspace and drop it next to the **Environmental Fact Table**.

2. Tableau Public automatically detects the field that occurs first in both data sources alphabetically and has the following properties:

   ○ The exact same field name, including capitalization and punctuation

   ○ The same field type

3. You can view and edit the **Join** details by clicking on the Venn diagram icon between the tables, as shown in the following screenshot. In this case, our data sources are joined by **Country Code**, which is correct. In order to select different fields, click on the name of the joined field and replace it with someone more appropriate:

4. When you are satisfied that the join condition is correct, click on the Venn diagram icon again.

 The additional fields, with their source table name appended in parentheses, are included in the data set below the workspace.

5. Just because a field is included in a join condition, it does not mean that you need to load it in the workbook. You also do not need duplicates of existing fields. For that reason, click on the **Country Name (Country Metadata)** and **Country Code (Country Metadata)** fields, which already occur in the **fact** table, and from their context menus, select **Hide**, as shown in the following screenshot:

6. If you would like to see the fields that you have hidden, click on the checkbox next to the **Show Hidden Fields** text above the data source.

The data source is now ready to be used in a visualization. To create a visualization, click on a sheet number or name in the ribbon at the bottom. If you'd like to get back to the data source, you can click on the data source icon from any worksheet.

When you load a new data source, the following are some of the several items that you should check before you can use it:

- Confirm that all the dimensions, are in fact in the **Dimensions** pane, rather than in the **Measures** pane

- Confirm that the data source types of all the fields are correct. For instance, in this data source, Year is formatted as a string, but it really should be a number

- We can change the data type by right-clicking on the field, selecting **Change Data Type**, and choosing **Number (Whole)**

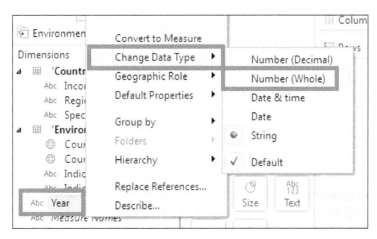

Connecting to web-based data sources

The steps required to connect to OData are different from the steps required to connect to the previously mentioned sources because they involve web servers and network security. These steps are a subset of the steps in Desktop Professional that are used to connect to a server:

1. Enter the URL of the website.

2. Select the authentication method.

3. Establish the connection.

4. Name the data source.

Another big difference between using local sources and online sources is that while the local sources can be refreshed with just a right-click on the data source name, online sources must connect to the website, which needs to be refreshed.

In order to refresh a web-based data source, perform the following steps:

1. Right-click on the data source name in the data pane.
2. Click on **Edit Connection**.
3. In the previous dialog box, which will be populated with the connection parameters, click on the **Connect** button in step 3 of the preceding list. It isn't necessary to repopulate the connection parameters or create a new connection to refresh the data.

Check out the visualization in the following screenshot, where we used this data source to graph the average CO2 emissions per capita by region since 1980:

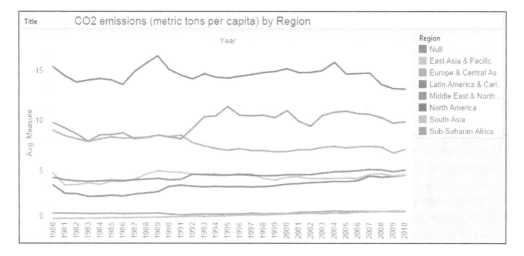

Summary

In this chapter, you learned one of the most critical concepts of using Tableau Public and producing accurate, informative data visualizations—connecting to, transforming, and loading data. The precepts that we discussed, such as making sure that your data source is clean and formatted properly, are relevant to working with data.

In the next chapter, we will discuss the different types of visualizations that you can make with your data and how to use each type correctly and in the most compelling manner.

4

Visualization – Tips and Types

A journalist or blogger who uses Tableau Public must convert data to visualizations so that readers can quickly understand large quantities of data that have been distilled down to a single graphic. The role of an author is data abstraction, the conversion of real-life data into visual cues (primitives, as some people who work in the field of data science call them), such as colors, shapes, lengths, and positions. The role of your reader is to do the reverse, consume data primitives and convert them back into real-life things, events, and phenomenon. If you have done a poor job choosing the elements of your data visualization, your readers will have a difficult time relating your visualization to the real world, which means that the story that you're telling isn't as compelling as it should be. Your visualization will not have the value that you wanted, because the connection between the data visualization and the underlying data has been broken.

Before you create a visualization for people to utilize, there are several considerations that go into the design. The following are some of these considerations:

- The purpose of the visualization: Which questions should it answer?
- Content: What should be included? (Start small, as less is often more).
- The visualization structure: What do you actually want to measure or show?
- The readers' platform: How have your readers approached your questions before? Which tools have they used to do so?

In this chapter, we will discuss the following topics:

- The lifecycle of visualization
- Iterative design and what it means
- Creating visualizations and visual perceptions

- Types of visualization
- Using discrete and continuous dates and measures
- Filtering, grouping, and sorting

An overview of the development lifecycle

Perform the following steps to develop a high-quality, compelling visualization:

- The first step is to devise a purpose for the visualization—a question that you would like the data to answer—and envision the possible visualization types that can best answer that question.

- Next, define the content and find your relevant data; determine how much data should be shown and the level of details required. A large part of your visualization project should be spent on finding and cleansing data. Do not skimp on this step, because it will make your visualization more clear and useful to readers. You may have to scrape data from the Internet or combine it from multiple sources.

- Wherever you get your data, make sure that you have both validated and attributed it properly.

 We surveyed a few experts in Tableau Public, and anecdotally, they estimate that they spend 70% to 80% of their total development time building data sources.

- The third step (with respect to this book) is to try various view chart types using the **Show Me** feature of Tableau Public. Try various view types to answer the question, experiment with various views, add or subtract data, manipulate components, and explore the question in greater detail or find other questions.

- The next step in building a visualization is considering the platform—print, desktop, laptop, tablet, and smartphone. The medium that your consumers use to interact with your visualizations and develop insights is critical.

- The last step is publishing your work, or delivering content to your visualization consumers. Your data visualization should start a conversation. So expect feedback.

Ten visualization tips

Data visualization is an art and a science. Much has been researched, studied, and published on the topic of what a great visualization is composed of. In this section, let's attempt to clarify some of these findings via some brief explanations, as shown in following list. As with everything in life, your mileage may vary, and there are always exceptions to rules. However, consider these general rules as condensed, combined, and paraphrased here as compared to the industry standards and academic research as you develop visualizations:

- Readers should be able to understand data visualization quickly. Make it easy for your consumer to read your data visualization. Don't use font sizes that are less than 10 pt if you can manage it, and use common fonts that are designed for ease of use. The key is that visualizations are not too flashy, complex, or artistic just for the sake of it. Do not use 3D in any way, it distracts and can mislead readers by distorting data. Follow the form over function principle, and the amount of time required to understand the data will work itself out.

- Consider the format for the reader. Print, computer screens, and mobile devices use large fonts for titles, and consider the format of consumers' devices. Scrolling is not good in most cases, and you should manage white space carefully.

- Every element should have a reason for being in a virtualization. Choose colors carefully. Use colors for categories instead of quantities; don't use colors randomly. Shapes, colors, legends, and labels should add to the understanding of the visualization.

- Be consistent in the placement of objects, elements, and colors. Your work is your digital brand, it literally is your product. Your consumers may not know you in person. Their perception of you and your skills is determined by how they relate to your visualizations. Do not use variety for variety's sake. The organization of elements and colors is important, as it helps readers navigate through the visualization.

- Use the correct context for data and consistent axes for elements on a chart. Help readers understand the data and what they are looking at. Do not mislead readers by using mismatched axes for charts that compare the same measure, nonzero axes, and other misleading tactics.

- Simplicity comes from clarity, and this helps readers understand data. Keep it simple! Some data sets are inherently complex. In such cases, try to show a part of the data set, or break up the visualizations to show the various aspects of the data. Be clear in your organization and approach, and simplicity (for readers) will follow.

- Consider calculating differences for readers instead of plotting multiple lines or bars in an effort to show the comparisons and differences. For example, a line chart of the city budget versus the city expenditures may be interesting, but an even better option is to use a bar chart, with a baseline set to the budget and bars above or under the baseline. Alternatively, you can set reference lines in a chart by using mean or target values.

- Choose the best type of view (chart) for the data. Read the following sections and understand the key elements of every major chart type and what the main objectives of each chart type are.

- Understand the different visualization primitives (color, length, area, shapes, position, direction, and angle) and how humans perceive them in terms of data visualization.

- Experiment with the Tableau Public **Show Me** feature to choose a view type and make changes of your own, such as selecting the shelf items that you need to include, and their characteristics. Experiment and refine, and think in terms of a **visualization cycle** of exploration rather than a linear route from data to visualization creation.

The perception of visual clues

Our brains interpret visual signals in specific, predictable ways, and modern user interfaces and consumer products are based on extensive study in this area. The way we interact with visual signals was first formally studied and discussed by *William Cleveland* and *Robert McGill* (at the time, with AT&T Laboratories) in their seminal paper named *Graphical Perception: Theory, Experimentation, and Application to the Development of Graphical Methods*, as published in the *Journal of the American Statistical Association* in September 1984. This paper is readily available via a quick Internet search, and it is worth a quick read.

In addition to *Cleveland* and *McGill*, who were statistical scientists (and later professors) focused on researching how we process visual cues, *Stephen Few* has written extensively on this topic in his books and on his blog, which can be viewed by visiting `http://www.perceptualedge.com/`.

The authorities on this matter have concluded that the human brain correctly perceives numerical values in a way that is different from how it perceives various visual primitives.

In the descending order of accuracy, visual primitives are as follows:

- **Position**: This shows how marks relate to an axis, like in a **scatter plot**.

- **The bar or line length**: This is why line graphs and bar graphs remain the first and the most popular forms of data visualization.

- **Angle**: This should be used sparingly, as the human brain actually is not good at differentiating angles within a circle. **Pie charts** are effective if you wish to communicate measures as long as about five dimensions are included and the slides are labeled with their respective percentages of the whole.

- **Area**: It's actually fairly difficult for us to differentiate numerical values from each other based on the size of a shape. Therefore, you should use bubble graphs and tree maps sparingly and label them properly when you do use them.

- **Saturation and hue**: These are the worst primitives that you can use to encode numerical values because not only are a significant number of readers color-blind, but also they are used inconsistently and designed for perception rather than judging.

- **Color**: This is best used for categorization into groups rather than to measure quantity. However, you can use it effectively in **heat maps and highlight tables,** as long as you are using it properly and labeling the values.

Your mileage may vary, but consider the hierarchy of visual clue perception carefully. Some conservative data authors will try to stick to scatter plots, line charts, and bar charts (and their derivative chart types) almost exclusively and avoid most other charts such as pie charts, bubble charts, and heat maps.

Leveraging your understanding of how our brains interpret visual clues will help you select the proper chart type. Fortunately, there is a built-in assistant in Tableau Public called **Show Me** that will help you choose a chart type based on the data elements (measures and dimensions) that you select.

Using the Show Me tool to create charts

Tableau Public has a great tool embedded into the software called **Show Me** (or the **Show Me** button). This tool is an expression of Tableau Software's vision of self-service analysis because it allows largely nontechnical users to develop useful graphs and charts based on their data. The **Show Me** button is at the top right-hand side of the Tableau Public interface. Clicking on this button opens the **Show Me** dialog box (with 24 chart type icons in Tableau Public 9.x), as seen in the following screenshot:

The **Show Me** tool suggests graph types based on the view that the user has created and/or the fields that the user has selected from the **Data** window or the shelf areas.

Only the graph icons that fit the data types available or selected are enabled. For example, none of the geographic maps on icon row number 2 are enabled, because there is no geographic data in the data set used to make this screenshot.

The **Show Me** tool also tells you which additional field types you need to select in order to create a specific visualization, which is a great way to learn.

To use the **Show Me** tool, select one or more dimensions or measures from the **Data** window and then click on the **Show Me** button in the upper right of the Tableau window to see the suggestions in the **Show Me** dialog box. The following are some examples:

- Selecting a continuous date field will prompt the suggestion of a line graph

- Selecting a geographical field will prompt the selection of a map

> You can't auto-hide the **Show Me** dialog box once it is displayed and the user has made a selection from it. Click on the **Show Me** button again to hide the **Show Me** dialog box.

In addition to creating new visualizations using the **Show Me** card, you can modify the existing graph. When you select a new graph type from one of the highlighted buttons, Tableau Public will change your visualization for you, and by doing so, it might remove fields on the visualization that aren't part of the new one.

Show Me contains 24 chart types in Tableau Public 9.x, but many more advanced or customized charts can also be created by adding different data elements and tweaking the existing chart that you have built. Items that help users better understand the data include tool tips, trend lines, color indicators, row separators, and reference lines. We will explore these additions in detail later in the chapter.

Answering questions using Show Me chart types

Of the many chart types available in the **Show Me** tool in Tableau Public 9.x, there are several common chart types that are frequently used — tables, line charts, bar charts, geographic maps, and scatter charts. Pie charts are often used in popular culture, particularly in infographics, but you should use them sparingly and always label the slices with the percent of the total.

Charts and graphs exist to answer questions, and some charts can naturally answer certain types of questions better than other charts.

The following sections discuss in detail some important chart types offered in Tableau Public, but this is not a comprehensive list due to limitations pertaining to space in this book. In order to help you learn how to construct different visualizations, the screenshots in these sections include the placement of different fields on the shelves and cards of the workspace.

About dimensions and measures

In *Chapter 3*, *Connecting to Data*, we discussed dimensions and measures in detail. Let's bring up a few more points so that we understand the chart creation process a little better. Tableau Public separates data source into dimensions (qualitative fields) and measures (quantitative fields).

When you drag a field onto a worksheet, its new instance is called a **pill**. The dimensions and discrete fields will be blue, while continuous measures will be green pills. Each of these pills has a right-click on the shortcut context menu and can have icons on them that correspond to the Marks card shelves that they belong, to as well as the set and group icons (if they are part of a group or set).

When we use a dimension, Tableau Public creates column or row headers for a chart (view). On the contrary, measures typically create an axis in the chart in case the measure is classified as continuous, as described in the next section of this chapter. Measures can be aggregated (mathematical calculations such as summation, averaging, counting, and so on) based on the selected aggregation function (SUM, AVG, COUNT, and so on) for each item in the dimension used in the chart.

Tableau Public automatically determines whether a data field is a dimension or a measure, but sometimes, the software makes mistakes. For instance, if your data source has numeric unique identifiers, such as account numbers, then they will be grouped as measures by default, unless the field name contains the ID string. In this case, you can change a data element to either a dimension or measure by dragging it to the correct pane in the **Data** window (to the **Dimensions** pane or the **Measures** pane). It is more likely that you will have to convert measures into dimensions (because all numeric values are not measures). To convert a measure into a dimension, you can either drag it to the **Dimensions** pane or right-click on the measure and select **Convert to Dimension** from the shortcut context menu. If this conversion is needed for only one chart, you can locally convert the measure on a shelf in the same manner (by selecting the **Convert to Dimension** command from the shortcut context menu).

Continuous and discrete dimensions and measures

As learned in the previous section, all data elements are classified in Tableau Public as either a dimension or a measure. These dimensions and measures are further classified into continuous or discrete elements.

The data elements in the **Data** window and on the shelves in Tableau Public are either light green or light blue in color. The green color indicates that a data element is set to continuous, and the blue data elements are set to discrete elements. A green or blue outline will also appear when selecting the corresponding continuous or discrete data elements. When you are building charts in Tableau Public and adding dimensions and measures to the **Columns** and **Rows** shelves, the continuous data items (usually measures) will create axes, and the discrete elements will create row or column headers.

Discrete and continuous designations for data elements affect how graph elements are built and marks and axes are rendered. For example, a continuous date dimension used in a line chart will create a traditional, unbroken line chart. Using a discrete date dimension will create a line chart with segmented panes, that is, lines that are broken apart with each change in the date part. For instance, for a graph that shows years and months, the lines will break between each year. Using continuous measures on the **Color** shelf will result in a color gradient being used in the chart, with the hue or color in proportion corresponding to the value of the measure. If a discrete measurement is used on the **Color** shelf, the color will not be a gradient but separate colors (this is generally fine for dimensions on the **Color** shelf, but not for measures). Have a look at the following screenshot to see the difference between using a discrete color and a continuous color.

In this example, we're using a continuous measure to color each cell. Note that on the **Marks** card, **AVG(Measure)** is green. This means that it's continuous. **Region**, as compared to **AVG(Measure)**, is blue because it's a discrete dimension (check out the next screenshot):

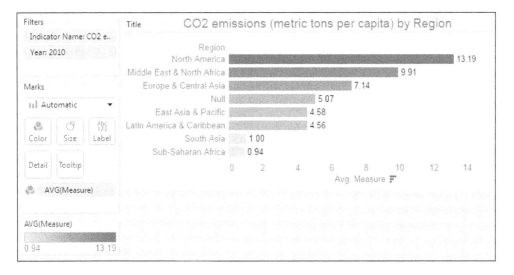

We converted **AVG(Measure)**, which is the average CO2 emissions per capita in 2010, into a discrete number by clicking on its context menu on the **Marks** card and then selecting **Discrete**, as shown in following screenshot. This version assigns a unique color to every numerical value:

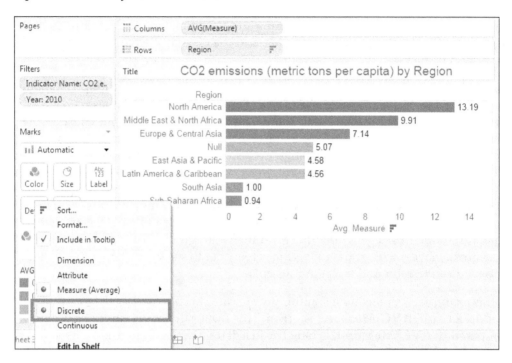

Similarly, if a continuous measure is used on the **Size** shelf, the sizes of the graph marks such as circles, squares, or other shapes, will be in proportion to the value of the measure. Discrete measures used in the **Size** shelf can have a more pronounced size variation, but you can experiment with changing the continuous or discrete settings for measures and dimensions to better understand their effect on the chart. In my opinion, using discrete measures on the **Size** shelf can sometimes yield good results.

Measures and dimensions can be converted from continuous elements to discrete and vice versa. This allows you to experiment and customize the chart view. Quite often, you will convert continuous elements to discrete elements, as you will also often convert measures to dimensions. You can convert continuous elements to discrete ones by right-clicking on the date element and selecting the **Convert to Discrete** menu option. Also, when you move a data element from the **Measures** pane in the **Data** window to the **Dimensions** pane, it will typically convert the date element from continuous to discrete.

Just like changing data items from measures to dimensions only in one chart, you can also change from continuous to discrete for only the desired chart instead of making the change globally for all the charts and worksheets in the workbook.

Selecting aggregation types for measures

Measures can be aggregated. These numeric values can be added, counted, and averaged, and the median can be chosen. In addition, the aggregation types that are available in Tableau Public also include statistical functions such as maximum and minimum selection, variance, standard deviation, and percentile. The most common aggregation types used in journalism and blogging are sum (addition), average, and count. The count aggregation is useful when you have a data source with rows that have an ID field, such as the FIPS ID, which is a numerical identifier for a specific geographic area, such as a county. To get a count of records, use Count(ID), where the ID is the specific ID name in your data source. Extending this example, you can also use **Count(Distinct)** when you only want a count of every unique ID and don't want to count the repeating IDs.

Selecting the aggregation type for measures can be done by right-clicking on a measure from the context menu, choosing the **Measure** command, and clicking on the desired aggregation type. You can also access aggregations from the drop-down menu by clicking on the tiny arrow on the right-hand side of the data element pill. The following is a screenshot of the full aggregation menu that is available from a measure pill's right-click on the shortcut context menu:

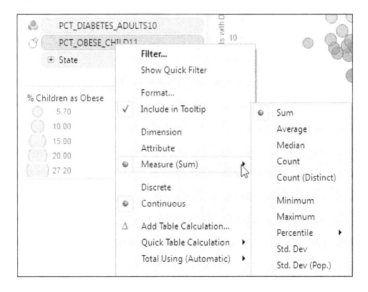

Swapping and sorting

Columns and rows can be swapped (one for another) by using the **Swap** button in the menu bar. Items on shelves can also be moved manually among shelves and to and from the **Data** window. Dimensions and measures can be duplicated by pressing *Ctrl* key and dragging to a new or the same shelf. This is helpful when changing aggregations for a duplicate date element or part of the date (such as changing year to month).

Similarly, you can sort fields in the following three ways:

- Right-click on the field in the **Data** window, click on **Default Properties**, and select **Sort**, which sets the default sort order but does not allow you to sort by a measure
- Click on the context menu of a pill on your visualization and select **Sort**
- Click on the ascending or descending sort buttons in the toolbar, or hover over a header and click on the appropriate icon

For instance, in the following screenshot, the countries in each region are sorted in the descending order by the CO2 emissions per capita. If you click on the quick sort icon that we have highlighted, the countries in each region will be sorted again in an alphabetical order. Conversely, when you hover over the x axis and click on the sort button, they will be sorted in the ascending order by the measure being used:

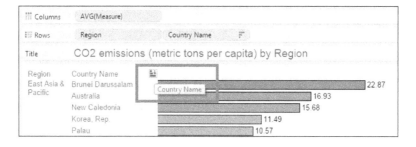

Types of visualization

There are many different types of visualization that you can create in Tableau Public, and we will focus on several of the most effective ones in detail. There are some that you should not use without practice, careful construction, and deliberate labels, such as bubble graphs, tree maps, and word clouds. We will not cover those here, since they have very limited uses.

Line graphs

Line graphs show data trends over time. Line charts, along with bar charts, are the most popular chart types used in data visualizations. For this type of time series chart, place a time or date dimension on the **Columns** shelf and a measure on the **Rows** shelf (the horizontal line chart is the most commonly used/useful line chart).

Line charts can then be modified and made more complex by adding a dimension on the **Color** shelf, which adds one or more additional lines to the chart. You can add different measures to the **Color** shelf as well and measures or dimensions on the **Size** shelf. But usually, adding too much complexity to the graph detracts from the understanding of the data. The following steps will help you create a line graph:

1. Create a line graph that shows the average military expenditures as a percentage of the GDP between 2000 and 2010 by loading the World Bank Indicators data.

2. Then, drag **Date** to the **Columns** shelf from the **Dimensions** pane.

3. Drag **Region** from the **Dimensions** pane to the **Color** shelf.

4. Drag **Military Expenditures (% GDP)** from the **Measures** pane to the **Rows** shelf.

5. Click on the context menu on the pill for **Military Expenditures (% GDP)** and select average the measure, as shown in the following screenshot:

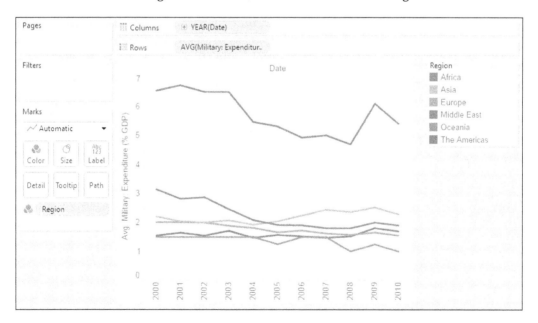

From this line graph, we learned that the Middle East has had the most fluctuation in its average military spending over the past 15 years.

Continuous versus discrete date-time elements

This section on line charts is a good place to discuss continuous versus discrete date-time elements. Line charts can be set up either with a continuous time series, or a discrete time series. Most line charts used for visualizations are continuous and unbroken, as the one in this section is. However, time can also be set as discrete, and this will break apart the time series into sections (also called panes or panels), such as by year, quarter, and month. Dashboards are sometimes developed with a discrete time series to allow for a fast comparison of quarters or months. A time-data element can be set as discrete or continuous by right-clicking on the date-time data element and setting it to continuous or discrete. Doing so will radically change the line chart by switching to either a continuous line or a broken line based on the parts of the date selected. Most journalists and bloggers who use Tableau Public will want to use the continuous feature of date-time data to create unbroken line graphs, but experiment with continuous and discrete to check whether it adds value for your readers.

Tables

Tables are usually created to show a fine level of detail for your data. In most business organizations, people have traditionally consumed analytics created in tabular reports, which can be highly inefficient. Tables display counts or measures relative to categorical variables, such as department spending, the number of college graduates in a city, pollution levels in a stream, and so on. They are useful when you wish to look up individual data point values and compare them across one or more levels of dimensional detail. Tables are the most effective when they are used at a high level and they contain data summaries rather than very long and detailed tables. You should also add filters to table charts to narrow down the data displayed to give readers a focus point. Tables are often called crosstabs. Pivot tables are a specific type of table; they are not discussed here.

There are three table views in Tableau — **text tables** (sometimes called crosstabs), **highlight tables**, and **heat maps**. Any of these three chart types can be selected and modified easily from one chart to another, or added to and varied as needed. The text table is a good place to start experimenting when building a table.

The tables on the US government's spending were obtained from Tableau Software. Download the Tableau Public workbook from Tableau by visiting http://tinyurl.com/1962-2012spending and download the source data from Tableau Software by visiting http://tinyurl.com/spending-source-data.

Text tables are common tables, as seen in excel, for instance. This type of chart is available for creation from the top row of the **Show Me** tool. The following screenshot is an example of a text table. This is a simple table that shows the US government's spending as a percentage of the total spending per **Department**, for the US **President**'s administration. For the table, we set the **SUM(Percent of Total Spending)** to the **Text** shelf, and this exposes the numeric values in the table rows and columns.

Note that the details of the spending (the numeric values) are easily seen, and there is no highlighting or context around the values (in terms of telling the reader whether the numbers are high or low). No shape or color marks are used. Also note that we set up various filters on the **Filter** shelf for this table view, including filters for only the top spending departments, and also limited to the last three presidential administrations (**Clinton**, **Bush**, and **Obama**). Because the data is only current through 2012 (and Obama started his first presidential term in 2009), we also set up a filter for **Administration Year** and filtered to show only years 1 to 4 corresponding to each president's 4-year term of office. This lets us compare equal numbers of years for each president:

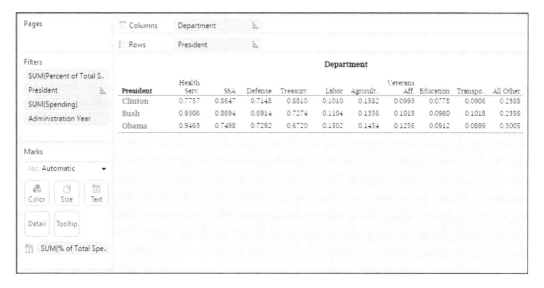

The heat maps are tables that are used to communicate visual cues such as shape and area by referring to up to two quantity measures. The larger the shape (and usually, the deeper the color), the higher the measure value. This type of chart is available for creation from the top row of the **Show Me** tool. The following graphic is an example of a heat map, with the **SUM(Spending)** on the **Color** shelf (in this case, the darker the red color gradient, the more dollars were spent) and the **SUM(% of Total Spending)** is tied to the **Size** shelf (this controls the size of the square). In this case, the larger the square, the more the percentage of the total spending for each government department during successive US presidential administrations. The actual values of spending are obscured here so that the reader is not able to quickly compare the actual values.

The colors used can be easily customized for each graph by double-clicking on the gradient color bar titled **SUM(Spending)**. The shape can be customized by selecting another shape in the dropdown list under the **Marks** title section. Filters that were similar to the **Text** table were set up. But this time, no filtering for presidents was created. All the presidents listed in the source data set are represented, from Kennedy to Obama, as shown in the following screenshot:

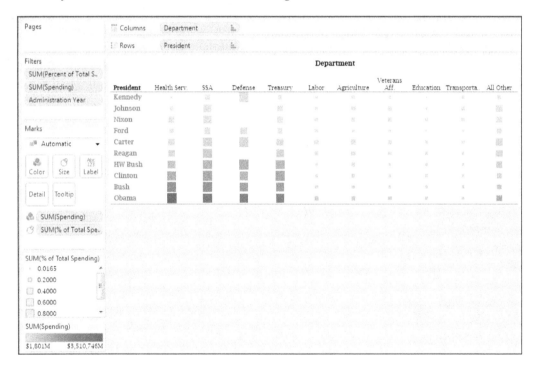

Highlight tables are often thought of as heat maps to the industry, as they use color gradient depth, saturation, or hue to highlight a quantity measure (such as highs and lows). The difference between a highlight table and a heat map is the lack of shape and size marks and the ability to see the underlying details in the highlight table. The following screenshot is an example of a highlight table. Note that the darker the color hue, the greater the percentage of spending to the total. The **SUM(% of Total Spending)** is duplicated in the **Marks** card for both the **Color** shelf and the **Text** shelf. This allows the color hue to be affected in proportion to the numeric value and also lets the value itself be shown in the rows and columns.

The colors used, and the other borders and elements, are completely customizable in the tables described in this section. The colors of this chart can be easily customized for each graph by double-clicking on the gradient color bar titled (in this case) **SUM(% of Total Spending).** The filters that have been created are identical to the first table named the **Text** table in this chapter section:

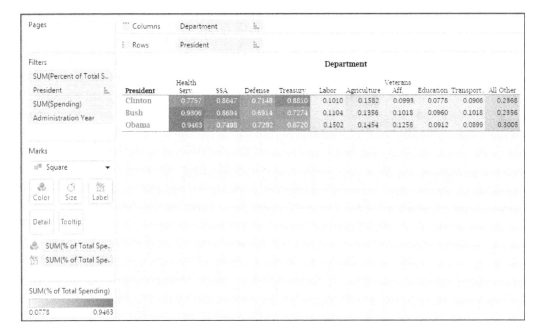

Bar charts

Bar charts are used to compare values and are perhaps the most useful of the chart types. Bar charts allow you to compare values across categorical dimensions (such as age groups, sex, race, cities, states, expense categories, departments, and so on; the list is endless), but they do not display the underlying data details as can be done in a table. Bar charts are used to compare values across dimensions. As we discussed in an earlier section in this chapter, length can be easily and naturally interpreted by readers. Therefore, bar charts are one of the more commonly used and appreciated charts used today.

> The global population data used in the following section was downloaded from the World Bank by visiting http://data.worldbank.org/data-catalog/Population-ranking-table.

There are several types of bar charts in Tableau — **horizontal bars**, **stacked bars** (they compare parts to a whole and are the same as pie charts), **side-by-side bars** (they compare two categories), Gantt charts, **histograms** (they measure the frequency of events and plot distribution diagrams), and **bullet graphs** (a relatively new compact chart invented by *Stephen Few*, a visualization expert; you should become familiar with his work). In this section, we will only talk about a couple of these bar charts, but I invite you to experiment on your own and learn about these other chart types if you find them appealing.

The following screenshot shows a horizontal bar chart that depicts the top 10 countries by population. The chart reveals that **China** and **India** are the two largest countries by population (with over 1 billion citizens each), followed by the USA, Indonesia, and other countries with a much lower population:

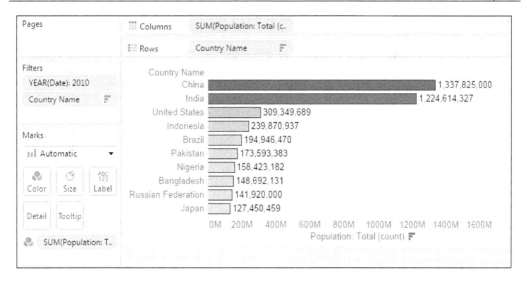

Some key design elements that should be pointed out in the previous horizontal bar chart are we used **Country** as the dimension and **SUM(Population)** as the measure. To limit the graph to the top 10 countries in terms of population, we added the **Country** field to the **Filters** shelf and chose to filter the top 10 countries by population in the descending order, as shown in the following screenshot:

After filtering the data to get information about the top 10 countries in the year 2010, we sorted the bars by clicking on the quick-sort button on the *x* axis to sort the countries in descending order by total population.

The chart shown in the following screenshot is a **stacked bar chart**. In this chart, we have pivoted several fields in the previous two exercises to show, by region, the 10 most populous countries in the world. The stacked bar chart is a **part-to-whole comparison** chart that often works better than a pie chart because it relies on the bar's lengths rather than an interpretation of the angles:

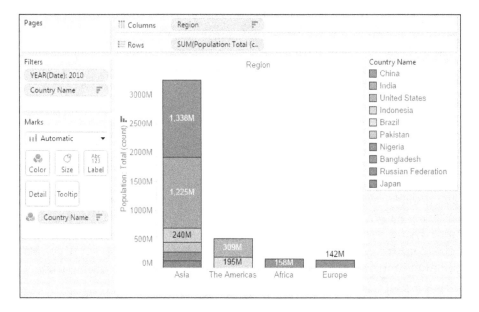

Stacked bar charts are best used when there are just a few different dimensions, because too many dimensions and corresponding colors are difficult to interpret. We created this stacked bar chart by performing the following steps:

1. Drag the **Region** field from the **Dimensions** pane to the **Columns** shelf.
2. Drag **Population: Total(count)** from the **Measures** pane to the **Rows** shelf.
3. Maintain the filters on **Year** for 2010 and the top 10 countries according to the total population.
4. Drag **Country** from the **Dimensions** pane to the **Color** shelf.
5. Click on the **Color** shelf on the **Marks** card and add a black border, which separates the bars more definitively.
6. Clicked on the **ABC** icon in the toolbar to see the mark labels.

Geographic maps

Geographic maps are a useful chart type that can answer various questions related to locations. Before you select this chart type, consider the fact that having geographic data elements such as longitude, latitude, state, county, city names, or addresses does not mean that you have to use them in every case. Sometimes, a bar chart or another chart type can express a data story as well as a geographic map in, case map data does not actually add value. For example, sometimes it is better to group states, counties, or cities into regions and treat those as categories by using bar charts.

Tableau Public has a powerful **geocoding** function that is built-in and requires no additional programming (though many third-party providers exist for more robust maps with *street* and *topographic* map support). Geocoding is the process of recognizing geographic data, such as addresses, cities, counties, state country names, latitude, longitude, and so on, with the help of software and automatically encoding text strings and numeric values as geographic dimensions.

We built the following map to show the maximum life expectancy by country in the World Bank Indicators data set by performing the following steps:

1. Multi-select **Country** in the **Dimensions** pane and **Life Expectancy at Birth (total)** in the **Measures** pane.

2. Click on the **Show Me** card.

3. Select the filled map.

4. On the **Marks** card, click on the context menu for **Life Expectancy at Birth (Total)** and change it to a maximum, which means that it shows the maximum value for each country within the data set.

5. Click on the context menu for the **Color Legend**, click on **Edit**, and select a color palette that's friendly to people who are color-blind, as shown in following screenshot:

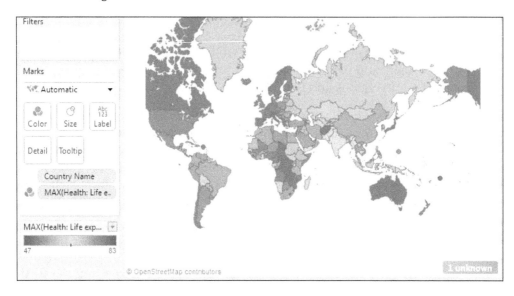

From this map, we learned that the countries of central Africa have relatively worst life expectancies in the world.

When the data is loaded in Tableau Public, the software automatically encodes geographic fields with their average latitude and longitude. When you select one of these newly encoded fields, a small globe icon will be associated with the data element. You can click on one of the map icons on the **Show Me** dialog box, or you can double-click on any of those fields to add it to the visualization pane. The filled map icon is the most commonly used one for data visualizations.

Scatter plots

Scatter plots are a type of chart that show relationships between two measures to establish correlations and comparisons, find trends in data, and expose outliers. Scatter plots are one of the most effective forms of communicating mathematical relationships, and they are one of the seven basic tools of quality control.

You can build a scatter plot by adding a measure to the **Rows** shelf and a measure to the **Columns** shelf, or double-clicking on each of the measures in succession — the first measure will be placed on the **Rows** shelf and the second one on the **Columns** shelf. You can swap the measures (rows swapped with columns) by pressing the *Ctrl* key and clicking on the measures and select the **Swap** menu command icon. Alternatively, you can right-click and click on **Swap** on the shortcut context menu. Then, add one or more dimensions to the **Marks** card, dragging various dimensions to the various shelves, such as **Color**, **Size**, or **Shape**. This will put the scattered measures in context with the dimensions and allow for an analysis of the relationships.

Aggregation of the measures plays an important role here. The default behavior of Tableau Public is to aggregate measures (for example. by using sums or averages). This may be fine for most analyses, but you may want a clearer picture (or you may want to expose outliers) by looking at all the data points. You can override this aggregation behavior by disaggregating the data, which will display all the values in the data source for that measure. To do so, go to the **Start** button and select **Analysis**. Then, click on **Aggregate Measures** and deselect that menu command. Note that disaggregation may remove the information displayed from the tool tip when hovering over a data point. You can add the contextual values displayed in the tool tip by dragging dimensions and measures from the **Data** window to the **Detail** or **Tooltip** shelves in the **Marks** card.

If too many values are displayed, you can add the measures to the **Filters** shelf and set parameters in the filter dialog box to limit the number of data points (marks) that are being shown.

We created the scatter plot shown in the following screenshot, which shows the relationship of GDP and mobile phone users between 2000 and 2010 in Asia by country from the World Bank Indicators data, by performing the following steps:

1. Drag **Finance: GDP (current. USD)** to the **Columns** shelf, which puts it on the x axis and makes it the independent variable.

2. Drag **Business: Mobile Phone Subscribers** from the **Measures** pane to the **Columns** shelf, which makes it the dependent variable.

3. Drag **Date** from the **Dimensions** pane to the **Color** shelf.

4. Click on the context menu for the **Color** legend and assign the **Cyclic** color palette.

5. Drag **Region** to the **Filters** shelf and select Asia.

6. Drag **Country** to the **Label** shelf.

7. Drag **Population: Urban (count)** to the **Size** shelf on the **Marks** card, which further encodes the default circles with the relative urban populations.

From this, we learned that the relationship between the **GDP** and **Mobile phone** subscribers in **China** is linear and that in other countries such as **Japan**, it is relatively consistent. It is also not surprising that the urban population of **China** is the largest, since it has the largest population. In the next chapter, you will learn how to create calculated fields that show you the percentage of the total that is urban rather than just a discrete number.

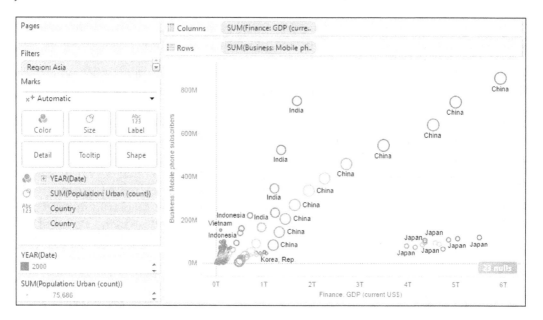

Pie charts

Pie charts allow you to show the composition of parts to a whole, like slices in a pie. Pie charts should be used sparingly because humans cannot interpret angles and area that well. Noted visualization experts, including *Stephen Few* and *Edward Tufte*, are generally against using pie charts, except in limited circumstances. It is often said for data, authors consider using a stacked bar chart instead.

That being said, pie charts can sometimes be used effectively under certain circumstances. Limit the number of slices to 5 (at the most), and make sure that the slices are not too small and are visible. You can use pie charts effectively to give a general sense of how one dimension compares to another, but don't use it to report measures that are close in value. It is too difficult for people to interpret when the measures (the pie slices) are of similar sizes. Tableau Public, unlike other data visualization software packages, does not support 3D pie charts and drill-downs into pie slices. This is a good thing because these characteristics lead to confusing and misleading pie charts.

We made the pie chart shown in the following screenshot, which still uses the World Bank Indicators data to show the relative percentage of mobile phone users in the world by region:

1. On a new worksheet, change the mark type to a pie on the **Marks** card.

2. Drag **Region** to the **Color** shelf.

3. Click on the context menu on the **Color** shelf to change the color palette to **Cyclic**.

4. Drag **Business: Mobile Phone Subscribers** to the **Size** shelf on the **Marks** card.

5. Click on the **Color** controller on the **Marks** card and add a black border.

6. Drag both **Region** and **Business: Mobile Phone Subscribers** to the **Label** shelf on the **Marks** card.

7. Click on the context menu for **Business: Mobile Phone Subscribers** on the **Label** shelf of the **Marks** card, select **Quick Table Calculation**, and choose **Percent of Total**.

8. Filter **Date** by setting it to 2010.

9. Click on the **Context** menu on the **Color** legend, select **Sort**, and sort **Regions** in the descending order by **Business: Mobile Phone Subscribers**:

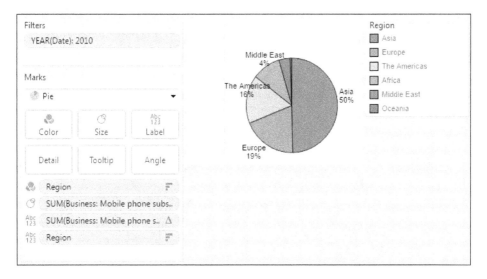

This pie chart shows that half of the mobile phone users in the world are in Asia. The pie chart is easy to use, and it does not mislead users. It provides them with the context that they need to sort through the information points quickly.

Using groups and sets

It can be useful to group dimension members into groups to report on these groups, as compared to others. For example, in the Life Expectancy Map that we built earlier, it might be helpful to add Canada, USA, and Mexico into a group or set named **North American Countries**.

A group is a simple set that is composed of the dimension members that you choose. In the following example, we set up the North American group by performing the following steps:

1. Right-click on the **Country** field in the **Dimensions** pane.
2. Select **Create** and then choose **Group** from the context menu.
3. Give the group a name.
4. Manually select the three countries that we want to put into the group.
5. Click on the **Group** button and then name the group.
6. Click on **OK**.

Note that the group field appears at the bottom of the **Dimensions** pane. It is considered to be metadata. It exists in the Tableau Public workbook, but it does not appear in your original data source.

In the following example, we added the group to the **Color** shelf on the **Marks** card of the map. Countries are colored by their membership in the group. Either they are in, or they are out, as shown in the following screenshot:

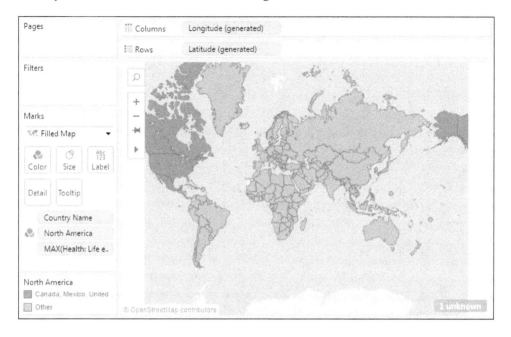

Similarly, we can create a set that groups members of a dimension by their adherence to the criteria that you establish.

We can create a set that identifies the top 25 countries by life expectancy, which will be used again to modify the map, from the World Bank indicators data source by performing the following steps:

1. Right-click on the **Country** field in the **Dimensions** pane.
2. Click on **Create** and then select **Set**.
3. Rename the **Set** to **Country Set**.
4. Click on the **Condition** tab rather than manually selecting the members of the **Set** so that it will be updated if and when we get new data.

5. Create a condition that includes the top 25 countries by **Health: Life Expectancy at Birth (total years)**.

6. Click on **OK**, as shown in the following screenshot:

We dragged the set from the bottom of the **Dimensions** pane and dropped it on the top of the field on the **Color** shelf on the **Marks** card, thus replacing it. Countries that are in the top 25 countries filtered according to life expectancy are colored blue, and all the others are gray, as shown in the following screenshot:

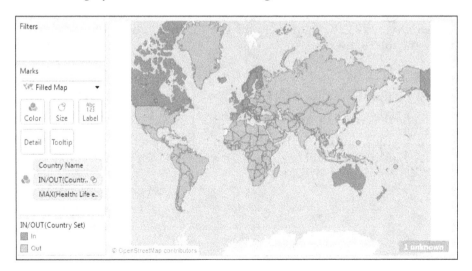

Summary

In this chapter, we discussed the processes that are the best if you want to design charts, from understanding the requirements, to data discovery, through to an iterative discovery cycle. We discussed the best practices in chart design and data visualization, reviewed the chart types that best answer which types of questions, and explored data dimensions and measures. We also discussed the different impacts that discrete and continuous data elements have on visualizations. Lastly, we talked about the use of reference and trend lines in views.

In the next chapter, we will talk about how to create calculated fields and table calculations.

5
Calculations

The data sources that you query to tell stories do not always have all the data points that you need. If they did, someone else would have told the story already. Tableau Public's data engine allows you to create new mathematical calculations of varying complexity, from basic multiplication functions to sophisticated aggregations with specific levels of detail that add color, context, and key insights to your data stories.

In this chapter, we will discuss how to develop new insights for users by understanding and implementing different types of calculated fields. We will use several data sources that we have connected to previously, in order to help you understand the following functions:

- Creating and editing calculated fields
- Types of calculations
- Number functions
- String functions
- Date functions
- Type conversions
- Aggregate functions
- Logic functions
- Blending data sources

Throughout this chapter, we will present and discuss error messages that you are likely to encounter. We cannot emphasize enough that, especially when presenting information to others using a tool that might be new to them, substance is more important than style. Also, it's very difficult to rebuild credibility once it's been compromised.

The next chapter focuses on Level of Detail calculations and table calculations. They both are big topics that build on the capabilities that you will learn in this chapter.

In this chapter, we will use a data set of major global floods since 1985, which was found on Tableau Public's resources page. The original source is at http://floodobservatory.colorado.edu/Archives/index.html.

Creating calculated fields

A calculated field that you create exists only in your data source in Tableau Public; it does not exist in the file that you originally queried, which is the reason behind why you are creating it in the first place. In the following examples, we will be creating calculated fields in a data set of floods. The fields that we create will show up in the **Dimensions** or **Measures** pane in the Data window in Tableau Public depending on their type of field, with the equals sign (=) next to them.

In Tableau Public, there are several ways to create a calculated field. The following steps will guide you:

1. From the **Analysis** menu, select **Create Calculated Field**.
2. Click on the **Context** menu of the data source and then select **Create Calculated Field**.
3. Right-click on a field in the **Dimensions** or **Measure** pane of a data source. Select **Create** and then choose **Calculated Field**.

Tableau Public has a new calculated field dialog box in version 9.x, which is a significant improvement over the previous versions. When it's open, you can perform functions on your visualization, and if you want to add fields to a calculation, you can drag them into the dialog box from the Data window.

You also can resize the dialog box and show or hide the list of functions.

The following screenshot shows the dialog box along with descriptions of each fields:

The following are the descriptions of all the functions shown in the preceding screenshot:

- **Calculated field name (1)**: This is where you give the field a good, descriptive name so that others understand what it is.

- **Data source (2)**: The data source name in which you created the calculated field.

- Click on the close icon **(3)** to close the dialog box.

- The list of functions **(4)** includes the following function types and elements:

  - List of function types, which can be used to filter the list of functions. The options, which will be explored in detail later in this chapter, are as follows:

    - Number
    - String
    - Date
    - Type Conversion
    - Logical
    - Aggregate
    - User
    - Table calculation
    - Filter box (here, you can type in the name of the function type that you would like to see)

- **Information box (5)**: Here, you can see the name of the function, what it does, and an example.

- **Button (6)**: This can be used to hide the function list.

- **The Apply and OK buttons (7)**: These buttons apply changes that you have made to the visualization and then close the window respectively.

- **Messages (8)**: This tells you when your calculation is valid, when there are errors, and what the error is.

- **The calculation itself (9)**: Tableau Public 9.x is intelligent. When you start typing in a field name, it will suggest all the fields that include the string that you are typing. You can hit the *Tab* key or click on its suggestion to populate the field. Alternatively, you can drag fields from the **Dimensions** and **Measures** pane, of both your current data source and others data sources that you have loaded into your workbook, to create new fields.

The parts of the formula include fields, mathematical operators, functions, and often parameters as shown in the following screenshot:

Editing calculated fields

Once you have created a calculated field that is valid, you can edit it by right-clicking on it in the window in which it appears and then selecting **Edit**. You can also rename calculated fields (as well as fields native to your data source) by right-clicking on them and selecting **Rename**.

Types of calculations

There are several types of calculations that you can create, as listed previously. Some of them are common to other data management tools, but some, such as user calculations, are unique to Tableau Public.

When creating calculated fields, the functions that you can apply on a field depends on the type of that field. For instance, you can't perform multiplication functions on a string field (even if it contains only numbers) without first converting it to a numeric field. Similarly, you can't concatenate numeric values without first converting them to strings.

Basic mathematical functions, such as addition (+), subtraction (-), multiplication (*), division (/), and exponents (^), are created by typing the standard operators into the dialog box.

You can add functions by typing their names or filtering the list of functions to the appropriate group, or name, and then double-clicking on the best choice. The different types of functions that you can perform in Tableau Public are as follows:

- **Number**: This includes geometric, trigonometric, and rounding functions, among others
- **Strings**: This includes options to find characters, measure string length, split and parse fields, and match character strings

- **Date**: This includes functions for duration, the addition and subtraction of dates, the truncation of dates, and the identification of date parts
- **Type conversion**: This allows you to convert fields into different types without modifying the metadata of the source field
- **Logical**: This includes the powerful IF, CASE, and ISNULL and allows you to tell Tableau Public how to group fields or relate them to parameters
- **Aggregate calculations**: These are the most commonly used functions; they include sum, count, avg, min, max, and median
- **User functions**: These are more commonly used for Tableau Server rather than Tableau Public; they allow you to create calculated fields that operate on usernames or groups
- **Table calculation**: This will be detailed in the next chapter; it allows you to compute aggregations based on data points that are visible in the current window

The number functions

Number functions include several functions that you may be familiar with owing to working with applications such as ROUND and ABS. Many of us may not have used several of these functions, such as the trigonometric and exponential functions, since our high school math class. There are some number functions that we will use extensively, such as MIN and MAX, while there are others that have limited uses, such as the geometric and trigonometric functions.

Number functions include the following, in order of decreasing precedence:

- **ABS**: This takes the absolute value of a number. It is commonly found in the denominator of table calculations.
- **CEILING**: This rounds up a decimal to its nearest integer and is the opposite of INT.
- **FLOOR**: This rounds a decimal down to its nearest integer.
- **MAX** and **MIN**: These take the maximum and minimum values in a sequence respectively. They are also considered to be string functions.
- **ROUND**: This allows you to specify how many decimals to round up a float number to.
- **ZN**: This is predominantly found in table calculations. It tells Tableau Public that if a numeric field has a NULL value, you want to replace it with zero.
- **DIV**: This produces the whole number product of a division statement.
- **PI**: This produces the numeric value of pi.

- **SIGN**: This produces a numeric value (1, 0, or 01) based on the input, which may be positive, zero, or negative respectively.

- **Trigonometric and geometric functions**: These are largely beyond the scope of this book. These functions include ACOS, ASIN, ATAN, ATAN2, COS, COT, DEGREES, RADIANS, SIN, and TAN as well as HEXBINX and HEXBINY.

- **Exponential and logarithmic functions**: These functions include EXP, LN, LOG, POWER, SQRT, and SQUARE.

The date functions

The date functions are useful for a variety of tasks, such as identifying the elapsed time between two events or a part of a date.

Many date functions in Tableau Public operate on dateparts. Dateparts are the small units of measurement of dates, such as a year, quarter, month, week, day, and the units of time, such as an hour, a minute, and a second. Writing date functions in Tableau Public is similar to writing date functions in ANSI SQL or Microsoft Excel with a big difference—dateparts need to be spelled out and enclosed in single quotation marks, which is a relatively simple design and use of punctuation for a programming language.

The most commonly performed analysis is the measurement of change over time, and the prevalence of date functions is high. Therefore, it is worth the time and effort to master them early on in your work with Tableau Public.

The date functions include the following functions:

- **DATEADD**: This adds numeric values, which can be hard-coded or variable, in a specific number of dateparts to a date.

- **DATEDIFF**: This calculates the elapsed number of specified dateparts between two dates.

- **DATENAME**: This produces the name of the specified datepart of a date. An example that demonstrates the implementation of this function is, `DATENAME('month', #7/21/2015#) = "July"`.

- **DATEPARSE**: This turns a string into a date in the format that you specify.

- **DATEPART**: This is similar to DATENAME. It returns the datepart of a specified date but returns a numeric value. An example that demonstrates the implementation of this function is, DATEPART('month', #7/21/2015#) = 7.

- **DATETRUNC**: This rounds up a date to the first date of the datepart that you specify. An example that demonstrates how to implement this function is, DATETRUNC('month, #7/21/2015#) = '7/1/2015'.

- **DAY**: This returns the day of the month of a given date. An example that demonstrates how to implement this function is, DAY(#7/21/2015#) = 21.

- **ISDATE**: This tests whether a string that you enter is actually a date and produces a Boolean value (true or false).

- **MONTH**: This produces the numerical month of a date.

- **TODAY** and **NOW**: These produce the date and datetime of the current moment by using the time settings on your computer.

- **YEAR**: This produces the year of the current date.

In the following example, we will use the floods data to determine the number of years that have elapsed since the last major flood. Most of the countries have had many floods and not just one. Therefore, we used the maximum value of **Date Began**, which is the most recent date in the data set, and compared it with today's date.

We could have omitted the **max** from the **Date Began**, but for each country, Tableau Public would have then aggregated the elapsed time per country, and we really wanted to know how long it has been since the most recent flood, as shown in the following screenshot:

Type conversions

We will begin with type conversions because you cannot create properly functioning calculated fields without having the fields in the right format. For instance, aggregations work on numeric fields, string functions work on strings, and date functions work on dates. Tableau Public automatically identifies the type of field based on its contents, but this may not meet your needs to aggregate or create calculations. Publicly available data sources, such as the ones available at `http://public.tableau.com/s/resources`, often have issues with either the format or the quality of the data. This will make it necessary for you to convert fields into different types, and it is quite likely that you will need to strip bad data and identify replacement values as well.

If you are familiar with relational database management, you are probably aware that there are many different field types, some of which are used to govern the field length within a database. Tableau Public focuses on the three primary field types, namely; string, which can be anything; numeric fields, which include integers and float fields; and datetime fields; which include the DATETIME, DATE, and TIME field formats.

You can convert a field to another type by creating a calculated field and wrapping it in the appropriate function. In case fields are of the same parent form or are interchangeable, you can right-click on the field name, select **Change Data Type**, and then choose the appropriate type. If you do need to change the format manually, you can use the following functions:

- **DATE**: This converts the strings that you specify (whether you enter them yourself or they are predefined) into date fields, which can then be used to drill through hierarchies.

- **DATETIME**: This converts a field or string into a datetime field, which is a date plus a timestamp.

- **FLOAT**: This converts integers into decimal numbers. This function works only on numeric formats.

- **INT**: This takes the whole number of a numeric expression by rounding up to the nearest whole number.

- **MAKEDATE**: This formats an entry for a year, month, and day into the MM/DD/YYYY format as a proper date. It also accepts fields for the datepart variables.

- **MAKEDATETIME** and **MAKETIME**: These functions accept proper dateparts, just like MAKEDATE, to convert variable or hard-coded sequences into the appropriate field types.

- **STR**: This turns a sequence of characters, numbers, letters, or special characters into strings.

The string functions

The string fields are often the richest in data source as they include other fields with free text, but they can also have the lowest data quality. The string functions in Tableau Public empower you to perform `splice`, `trim`, `find`, `replace`, `match`, `reformat`, and `concatenate` functions.

We will use several of these functions to clean up the floods data source, which was assembled manually and has several data quality issues.

The following are some of the most useful functions that we will use:

- **CONTAINS**: This has a Boolean output. It tests whether a field contains the specified string.

- **FIND**: This finds the place where a string of characters is located within a field.

- **LEFT**, **MID**, and **RIGHT**: These often work in conjunction with FIND and LEN when extracting fixed or variable strings of characters from a field.

- **LEN**: This produces the length of a field.

- **MIN** and **MAX**: These are commonly used on numeric fields. They produce the numerically minimum or maximum values in sequences respectively.

- The **REGEX** expressions: These features were introduced in Tableau Public 9.0. They extract, match, and replace variable strings within fields and are similar to the LIKE function in ANSI SQL.

- **REPLACE**: This replaces a sequence with a specified value.

- **TRIM**, **LTRIM**, and **RTRIM**: These trim leading or lagging spaces from a string.

- **UPPER** and **LOWER**: These are commonly used to normalize the contents of a field.

The Tableau Public does not have the PROPER function that both ANSI SQL and Microsoft Excel use to impose proper capitalization on inconsistently capitalized fields.

In the next few examples, as shown in the following screenshot, we will use the TRIM and REGEX functions to repair the quality of the most varied misspellings of several countries. It looks like many of the values in the **Country Name** source field have spaces on either side of the primary value, which is not great and it's something that we need to fix:

This is helpful. We now have fewer unique entries for country names than before, but bad characters are still present in the data-set.

There are two options for the replacement of characters in a string, namely REPLACE and REGEXP_REPLACE:

- **REPLACE**: This requires you to enter the string to be searched, the exact pattern to be replaced, and the replacement value. The advantage of using REPLACE is that it is easy to learn and execute. The disadvantage is that Tableau Public can search and replace only one string at a time.

- **REGEXP_REPLACE**: This is a robust function, and while it takes practice to master it, this is widely used in different programming languages. It is a type of regular expression. Among the technical users of Tableau Public, the introduction of regular expressions in 9.x was highly anticipated because the search pattern allows a high degree of variability. You can enter a specific letter, number, or special character as the pattern that you need to search, or you can tell Tableau Public to find letters, numbers, or a variety of special characters. You can also ask Tableau Public to find a combination of these.

Regular expressions are efficient, and they require less hard-coding than the REPLACE function. This means that you will need to modify them less as your data set changes. This is important with data sets that you (or other people) are compiling because humans inevitably introduce errors. Anyone can fat-finger a country name when they're entering data.

The regular expression that we entered includes the string to search (**Country source**), the pattern to search, which is entered in quotation marks, and the replacement, which is blank, as shown in the following screenshot:

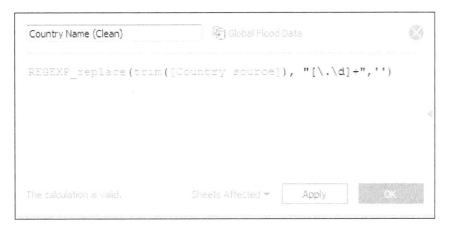

The pattern that we entered, which is also called a token, has several parts. This is shown in the following screenshot:

```
"[\.\d]+",
```

Everything within the quotation marks is a *token*. Working from the outside, the plus sign asks Tableau Public to match any of the expressions inside the brackets. Within the brackets, there are two expressions that need to be found, namely a period and a letter d. These expressions are very different from each other; the period is a literal, which means that we want Tableau Public to find all the periods. The d is a variable, and it means that we want Tableau Public to find numbers.

The two expressions, namely the period and the d variable, are preceded by a backslash. This is an escape. The concept of escapes is beyond the scope of this book. However, it is critical to master them if you want to learn how to structure data because they tell programming applications where to break apart long strings of variable characters, such as URLs. In this case, they ask Tableau Public to look for exactly the expression that we have entered.

If you would like to master the advanced string functions of Tableau Public, which are portable to other programming languages and are a good investment of time in case you're planning on producing advanced analytics or would like to build a career in data science, check out Mark Jackson's (`http://www.twitter.com/ugamarkj`) blog post at `http://ugamarkj.blogspot.com/2015/01/tableau-90-and-regular-expressions.html` and Joshua Milligan's (`http://www.twitter.com/vizpainter`) post at `http://vizpainter.com/my-favorite-tableau-9-0-feature/`. Mark and Joshua are both Tableau Zen Masters. It's a coincidence that the search pattern in our example matches that of Mark's. In line with full disclosure, Joshua Milligan is one of the reviewers of this book and an author who has worked with Packt Publishing on a book on Tableau as well.

The aggregate functions

Aggregation functions in Tableau Public are performed typically on numeric fields. In this section, we will show you how to use the default aggregations on visualizations as well as how to use them in calculated fields.

The following are the aggregate functions that are available for you to apply on a field are also available in the calculated field dialog box. Tableau Public has arranged them in order from the greatest to the least commonly used functions for the visualization as follows:

- **SUM**: This adds up the values within a partition
- **Average**: This sums the measure and divides it by the number of dimension members in the partition
- **Median**: This provides the measure value that's halfway between the greatest and least values
- **COUNT** and **COUNTD**: These count the number of dimension members and the number of distinct dimension members respectively; they are typically performed on dimensions rather than measures
- **Minimum** and **Maximum**: These take the least and greatest values in the partition respectively
- **Percentile**: This provides the numeric value at which a percentage of the partition that you specify falls beneath (we'll use this for several examples)
- **Standard Deviation**: This is the square root of the variance for a data set, and it is the unit of measurement of distance from the mean within a data set; it is commonly represented by the Greek letter, sigma

- **Variance**: This is the average distance from the mean value, and it's used to measure the distribution of a data set

The variance and standard deviation are two of the basic measurements in statistics. Advanced statistics in Tableau Public are beyond the scope of this book, but if you understand the concepts, you will find that there are several capabilities within Tableau Public that provide rich statistical functionality.

The most basic aggregation is a sum, and it's the default for numeric fields in Tableau. The second and third most basic aggregations are counts and averages. When you add a measure to a visualization, Tableau Public automatically sums it. In the following example, which uses the data on global floods that we referenced earlier, we do the following three things to create a map:

1. Double-click on **Country** in the **Dimensions** pane to create a symbol map.

2. Double-click on **SUM(Affected sq km)** in the **Measures** pane to change the map into a filled map.

3. Right-click on the **Context** menu for the color legend. Change it to a reversed *orange-blue diverging color spectrum* so that the higher numbers represent greater flood damage, as shown in the following screenshot:

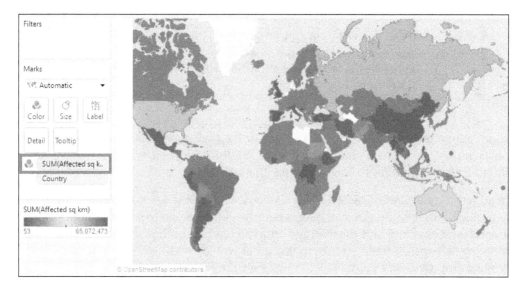

Note that on the **Marks** card, Affected sq km is summed. Therefore, you see the total area affected for each country in the whole data set. There is a major flaw with this method of aggregation, it is not normalized. It would be more useful to have a rate, such as the average area affected per flood, or to add context by creating an aggregation that tells you the percentage of the total area that was affected. For instance, we know that in Pakistan, there were catastrophic floods in 2010, and China traditionally also has had severe floods. However, these countries are in blue, which seemingly indicates that their floods were less severe.

Let's start changing aggregations, and thus add context and reduce the likelihood that someone will misinterpret our visualization, by changing the Measure function to an Average function. The easiest way to do this is by performing the following steps:

- Click on your aggregation context menu.
- Click on **Measure(SUM)**.
- Click on **Average**.

Changing this measure to an **Average** will average the square kilometers of all of the rows related to each country. This means that if there were three floods in the US and they affected 1 million, 2 million, and 3 million square kilometers respectively, then the average would be 2 million.

 If you would like to set the default aggregation for a field, you can right-click on it in the **Measures** pane, click on **Default Properties**, select **Aggregation**, and select the appropriate aggregation for your needs.

The result is a visualization that shows much more appropriate context. From this, we can gauge the average severity in terms of the total area affected by each flood in every country. In the next chapter, we will rank the countries, which adds even more context and relates them to each other for the user.

We added the number of records as well as the sum of the affected square miles to the tooltip so that when you roll over a country, you see the total number of floods, the total area affected, and the average area affected by each flood, as shown in the following screenshot. Using a tooltip is a great way of adding context, which is the art of relating two data points to each other and to the consumer:

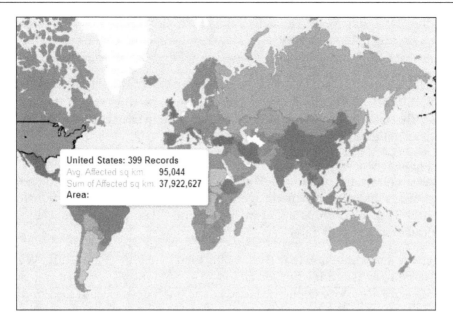

The logic functions

Logic functions enable you to tell Tableau Public what to do when certain conditions are met. They are also known as conditional statements. The format is commonly referred to as IF THEN ELSE. You, as a programmer, ask Tableau to test whether a row of data meets a certain condition. If it does meet the condition, then there is an output, which can be another field, a discrete number, or a string. If the condition is not met, then you want a different output.

There are several logic functions in Tableau. Some of these functions are sub-functions or parts of others. We will focus on the following major functions:

- **IF**, followed by **THEN, ELSE-IF**, or **ELSE**: This tests whether a condition is met and show the result if it is met as well as other conditions that need to be tested and results that need to be produced in case none of the conditions are met.

- The **IIF** statement: This tests for a minimum of one condition and then provides the results when the condition is or is not met. It follows the same format as that of IF in Microsoft Excel.

- The **IFNULL** function: This identifies what Tableau Public should do in case a value is null. This is particularly useful for aggregations of dirty data sets, which are very common.

- The **CASE** statement: This tells Tableau Public what to do when a parameter or a string field has a very specific value. Unlike the CASE statements in ANSI SQL, Tableau Public does not accept aggregations in these.

- **AND**, **OR**, and **NOT**: These link or negate conditions that need to be met.

- **END**: This is critically important, as it terminates the loops of the IF and CASE statements. Tableau Public will tell you in case you need to add it and have not done so.

The IF statements are very useful. You can use them to group members and set thresholds, among many other uses. The condition that needs to be met can be a variety of things. In the first example, we will create a new measure in a transformation of the World Bank indicator data that we have extracted and transformed, and it's called population and land data.

We will create a new measure that gives us the total land area of a country. We want to use this measure to determine the percentage of each country's area that was flooded, but this measure does not exist in the Floods data, but it does exist in the **Population and Land Data**. Later in this chapter, we will blend the data sources to create a calculated field that uses it in the denominator.

The Population and Land Data does not have a metric called total area. The way the data source is structured, the measures are in rows and not columns, with a field representing the corresponding metric value, as shown in the following screenshot. We need to create a calculated field that states that if **Indicator Name** is **Land area (sq. km)**, then we want the **Indicator Value** to appear as follows:

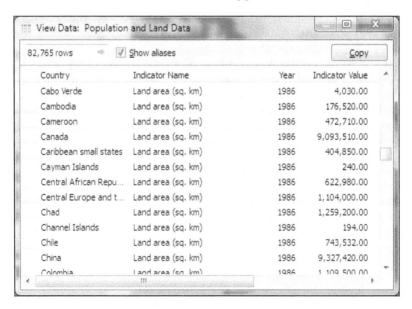

In order to do this, we will follow these steps:

1. Create a new calculated field in the **Population and Land Data** source by clicking on the **Context** menu in the Data window and selecting **Create Calculated Field**.

2. Name the calculated field *Area of Country*.

3. Start an IF statement that says that if the **Indicator Name** is **Land area (sq. km)**, then Tableau Public should use **Indicator Value**. Otherwise, it should use 0.

4. End the statement.

5. Validate that there are no error messages.

6. Click on **OK**. The result is shown in the following screenshot:

In this statement, if we did not include the ELSE statement, then the products of the field would be null for every value of **Indicator Value** that isn't specified, and we do not want nulls, because performing any kind of mathematical operation on a null value results in a null value even when you're adding or multiplying it by valid numbers. So, we included **else 0**, even though we did not necessarily need to do so, in order to preserve the integrity of the data set.

Another good use of an IF statement is to create groups or establish performance thresholds. In the World Bank Indicators data source, there is a field that identifies the percentage of its GDP that each country spends on public health. The first quartile is at five percent, the median is at seven percent, and the upper quartile is at nine percent. In the following example, we will use mathematical functions to group the countries according to what their average expenditure on public health is. (You can download this data set from www.worldbank.org).

In this example, we have hard-coded the values in the inequalities, which means that the thresholds used for the groups are not dynamic. In the next chapter, we will use window calculations to make them dynamic.

We created a simple box plot that shows the maximum health expenditure as a percentage of the GDP per country in 2010. Since we put **Country** on the **Detail** shelf, Tableau Public graphs one mark per country. In the calculated field, we even accounted for null values, as represented in the ELSE statement. Then, we put this new calculated field on the **Color** shelf of the box plot, as shown in the following screenshot:

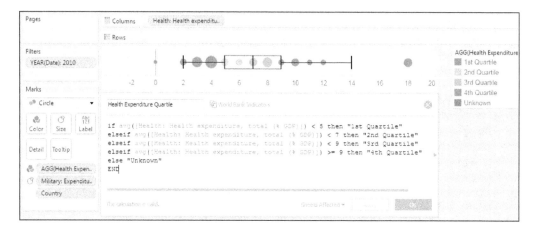

Perform the following steps to create this graph:

1. Add **Health: Health expenditure, total (% GDP)** to the **Columns** shelf.
2. From the **Analysis** menu, deselect **Aggregate Measures**.
3. Drag **Date** to the **Filters** shelf and select **YEAR**.
4. Filter **YEAR** to 2010.
5. Drag **Military: Expenditures (% GDP)** to the **Size** shelf.
6. Drag **Country** to the **Detail** shelf on the **Marks** card.
7. Create a new calculated field called **Health Expenditure Quartile**, and provide Tableau Public with several conditions to test.

We wrote the conditions in a specific order so that for each country, the accurate value is the output.

8.   Drag the new field named **Health Expenditure Quartile** to the **Color** shelf.

9.   Adjust the colors so that the lower quartiles are in red and the upper quartiles are in blue.

This IF statement is similar to the two other types of logic statements in Tableau Public, namely IIF and CASE.

The structure of IIF is the same as that of the IF statements in excel, and it's very similar to the IF statement that we just created in that you are asking Tableau Public to test a condition and then provide the results according to the logical outcome, which can be either true or false. In the following example, we will ask Tableau Public whether the last letter of the name of each country is *A*.

There are two possible outcomes — yes, the condition is met, and no, it is not met. We tell Tableau Public what value to use for each possible outcome. We can use a string, a number, another field or aggregation, or even nothing at all.

In the following example, which is for demonstration purposes, we ask Tableau Public to test whether each country name in the global major floods data, which was trimmed in the section on string functions, ends with the letter A. If the condition is met, the result should be blank (which is different from NULL), but in case it is not met, then the outcome should be the last letter of the country:

```
Last Letter                    Global Flood Data

IIF(right(rtrim([Country Name]),1) = 'A',
'',
upper(right(rtrim([Country Name]),1))
)



The calculation is valid.    Sheets Affected ▾           OK
```

Blending data sources

If you want to create a calculated field that produces the percentage of the total land flooded during an event, we need to use a data source that has this information in it. Most stories need more than one data source. When you are adding in secondary data sources, it's important to know what the level of aggregation is if you have joined data sources with different levels of aggregation and have not accounted for that when blending them because if you fail to do so, you might get the wrong results in your calculated fields.

The flood data is aggregated at the event level. This means that for every major flood, we have summarized data—the start date, end date, location, total land area affected, and so on. This is moderately granular data. If it were even more summarized, it might be summed up as floods per year per country.

We want to know the total percent of land flooded each year, and for this, we will assume that the area of each country is static throughout time. So, we created a new data source. We filtered the World Bank indicators data (available at `http://data.worldbank.org/data-catalog/world-development-indicators/`) in this data source to include only the indicators pertaining to the land area or population. The World Bank indicator data is aggregated by country by year, which is not the same level of aggregation as that of the data on floods; it is higher. So, in our calculated field, which will sum the total area flooded and then divide it by the total area of each country, we will need to tell Tableau how to aggregate the total area of each country. Otherwise, we will get the wrong results.

First, we need to join the data sources. In the companion workbook, which can be found at [X], we have a data source called **Population and Land Data**. This is the version of the World Bank indicators that we have filtered. It has only four dimensions, in this case, we have three, **Country**, **Indicator Name**, and **Year**. The data source on floods has Country Name and several other date fields. We want to join on **Country** name, as shown in the following screenshot:

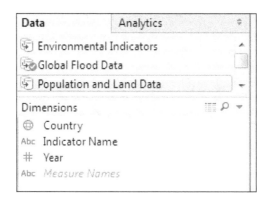

If we look at the **Population and Land Data** source, we do not see any of the gray chain links next to our dimensions that indicate that the fields are linked. There's a reason behind this. The names of the fields are not exactly the same. In order for Tableau Public to identify the join conditions automatically, the field names need to be exactly the same, with both capitalization, punctuation, and content. The field types need to be exactly the same as well.

From the **Data** menu, we can create a join on **Country** name in the Floods, as follows:

1. Click on **Edit Relationships**.

2. In the **Edit Relationships** dialog box, which is shown in the following screenshot, ensure that the appropriate data source is selected as the primary data source, which is the data source from which the first field is added to the visualization.

3. Select the appropriate secondary data source.

4. Click on the **Custom** radio button in case it is not selected by default. In this case, there are no automatic joins. Therefore, Tableau Public assumes that we need custom joins.

5. Click on **Add**:

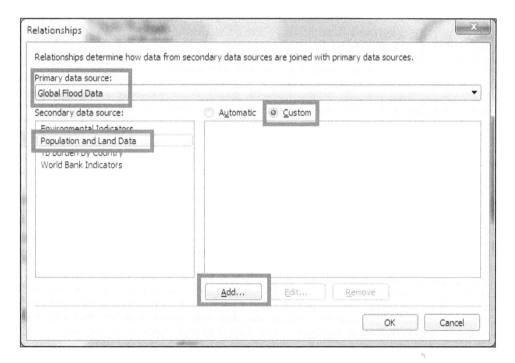

Next, we need to tell Tableau Public that the Country Name field in the major global floods data, which was created to remediate data quality issues, is the same as the Country field in the major global floods data source, as shown in the following screenshot:

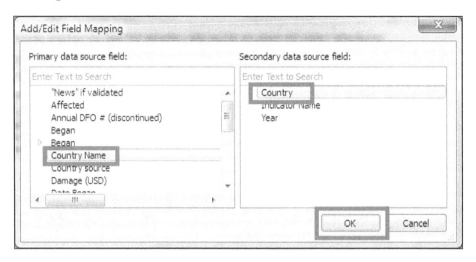

You can create only one custom join at a time. So for instance, if you would like to add more joins, such as a date field, you need to add each one individually by clicking on the **OK** button and then clicking on the **Add** button again, which we did in order to add a second join on the year of the event.

Note the small arrow next to the **Began** field, which is above **Country Name**. This field allows you to establish a join condition on a specific date part. Since we also wanted to join the data sources on the year of the flood event, we expanded the list to see the appropriate dateparts, clicked on **Year**, and then selected **Year** in the secondary data source.

We clicked on **OK** again to return to the visualization, which is still the map that we created that shows the average flooded land.

The view is now different. Looking at the **Data** window for the **Population and Land Data** source, you will see that that there are two chain links, the chain link for **Country**, and the following is orange. Because the linking field is a part of the visualization, you will see that **Country Name** is on the **Detail** shelf on the **Marks** card. However, **Year** has a gray, broken chain link because it is not a part of the visualization. If you want it to be used in the calculated fields, click on the gray chain link to activate the join. Even though the field might not be on the visualization, it still be a part of calculations using fields from its data source, as shown in the following screenshot:

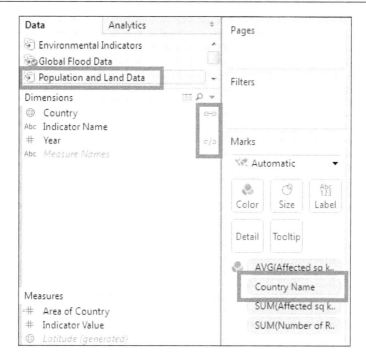

We can create a calculated field that produces the average amount of land in each country that was flooded. In this case, since some countries might have had major floods that add up to more than 100 percent of the total flooded land, we will create the maximum percent of land flooded, as shown in the following screenshot:

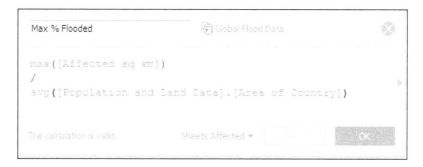

In order to create this calculation, perform the following steps:

1. Create a new calculated field in the major global floods data source.

 It is important that the place where we create the new calculated field determines the data source that is the primary data source.

2. Name the field.

3. Add **Affected sq km** to the numerator by dragging it into the dialog box.

4. Wrap it in a **max** function.

5. Add a division sign (/) because we want to divide it by the total area of the country.

6. Click on the **Population and Land Data** source and drag **Area of Country** to the denominator.

7. The denominator is automatically summed. You can change the sum to whatever aggregation you would like, but if you do not aggregate the fields that are from the secondary data sources, you will get error messages.

8. Change the aggregation of the field from the secondary data source, which is the field whose name is preceded by its data source name. This is similar to a SQL query that is used to find out an average. The reason behind why we did this is that for every record in the partition, Tableau Public needs to know what to do with the denominator. If we sum it, we will not have the proper calculation.

Check out the correct calculated field, It's important to learn how to handle errors. The next two errors are very common and are as follows:

- The following screenshot shows that we cannot mix aggregate and non-aggregate arguments with this function, with the division function highlighted. This means that in a calculated field, both arguments must be aggregated. We can resolve this by aggregating the numerator or by using the ATTR function on it. ATTR tells Tableau to use the discrete value for it:

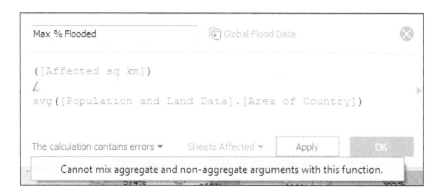

- All fields must be aggregate or constant when using table calculation functions or fields from multiple data sources. This means that the denominator, which is from a secondary data source, must be aggregated. Tableau Public requires this so that it knows how to aggregate. We resolved this by placing the AVG function around the denominator and adding an aggregation to the denominator:

Summary

In this chapter, we explored the different data types in Tableau Public. You learned how to create various types of calculated fields, which enable you to create new metrics and tell stories that are relevant and impactful to users.

In the next chapter, we will build on these capabilities to create Level of Detail calculations and table calculations.

6

Level of Detail and Table Calculations

In the previous chapter, you learned how to use custom calculations to create new fields that add context and insights to the story that you are telling with Tableau Public. In this chapter, we will discuss how you can use table calculations and level of detail calculations to enhance the comparisons that you are making with the data. We will build on several of the calculations that we have created to show you how to make them more dynamic and contextual.

The table calculations are different from the traditional calculations because they are performed locally on the data in the cache, that is, the data that Tableau Public has used to create the visualization. Not all data that is in your data set is in use all the time. The data in the cache is what Tableau Public is using in the memory to render your visualization. Therefore, the data that you have filtered out will not be included in the table calculations.

Level of detail calculations, which are a new feature in Tableau 9.x, allow you to tell Tableau Public at exactly which dimensional level you want the calculations to be aggregated. We are discussing them along with table calculations because some of the concepts are very similar.

Table calculations have the following major functions:

- Calculating change over time as a relative percentage
- Calculating the percentage of a whole attributed to a one-dimensional member
- Calculating change relative to other members of a dimension
- Distinguishing the maximum or minimum values in a partition
- Moving calculations, such as averages and sums

- Running calculations, such as running sums, that are used for waterfall and Pareto graphs

The following are the level of detail calculations that Tableau Public enables you to do:

- Fix the level of aggregation
- Include specific dimensions, that may not be present on the visualization, in calculations
- Exclude specific dimensions, that are present on the visualization, from specific calculations

In this chapter, we will discuss the following table calculations:

- Creating quick table calculations
- Addressing and partitioning table calculations
- Changing over time
- Editing table calculations
- Moving averages – window max and running max
- Ranks and percentiles
- Difference from the average

The level of detail calculation concepts that we will discuss involve fixing, including, and excluding dimensions, as well as nesting calculations and limitations. A brief exercise on editing fields in the shelf is also included in this chapter.

About data sources

We will continue using the World Development Indicators data source, as some of these examples will be included in the dashboard that we will develop later in this book. You can download the data from the World Bank by visiting `http://data.worldbank.org/products/wdi`.

Creating quick table calculations

Tableau Public has a feature that enables the rapid creation of a table calculation. After dragging a field onto the visualization, typically a measure, you can aggregate and perform table calculations on dimensions, or click on its context menu and select **Quick Table Calculation**. There is one limitation to this—table calculations cannot be created on fields that have forecasting turned on.

The following are some of the many different types of quick table calculations, though not all of them are available at the same time, and the options vary by the type of dimensions and measures that are also on the visualization:

- Running total
- Difference
- Percent difference
- Percent of total
- Rank
- Percentile
- Moving average
- Year to date (YTD) total
- Compound growth rate
- Year over year growth
- Year to date (YTD) growth

Once you have created a table calculation, you can edit it by either clicking on the **Context** menu on the pill, or right-clicking on it. For each unique table calculation, we will explain how to modify it and what the components of the formula mean.

Changing over time

One of the questions that we commonly ask about data is, how has performance changed over time? This helps us understand how events of decisions affect outcomes. While understanding the relationship between discrete numbers is helpful, it's even more useful to understand the rate of change, or the percentage change over time. The reason why it's important to understand the rate of change is that, while discrete numbers may appear to be increasing, the actual percentage change from year to year may be declining.

In the following example, we will discuss how to graph the amount of remittances over time and we will create a quick table calculation that shows the percent difference:

1. On a new worksheet, using the World Development Indicators data source, drag **Year** from the **Dimensions** pane to the **Columns** shelf.
2. Drag **Remittances (USD)** from the **Measures** pane to the **Rows** shelf.
3. Exclude years before 1980 by using your preferred method. We selected them on the *x* axis, hovered over pill, and selected **Exclude**.

4. We now have a basic line graph that shows change over time, but we're missing context. While we can see that **Remittances** have gone up and down, we know nothing about how they relate to the economy of the region or its population.

5. We can edit the measure to produce a rate per capita. We will do this by editing it in the shelf, which is a new feature in Tableau 9.x, and adding the total population to the denominator (editing in the shelf is a good option for simpler aggregations).

6. On the **Rows** shelf, click on the pill for the measure to see the **Context** menu and select **Edit in Shelf**.

7. Enter a division sign after the closing parenthesis, and from the **Measures** pane, drag **Total Population** to the right of the division sign so that it is in the denominator, as shown in the following screenshot:

Rows   SUM([Remittances (USD)])/sum([Total Population])

Apply (Ctrl+Enter)

Since the **Remittances (USD)** is summed, we need to sum **Total Population**. In a calculated field, both fields must be aggregated or disaggregated. We could disaggregate, but that would give us the average per country within each region, and when those numbers are added up, they will not represent the data accurately.

8. In order to give the new field a name, drag it from the **Rows** shelf to the **Measures** pane, which does not seem intuitive. When prompted, name it **Remittances per Capita**.

9. We can create a quick table calculation that shows change over time by right-clicking on the **Remittances per Capita** field on the **Label** shelf, selecting **Quick Table Calculation** from the context menu, and then selecting **Percent Difference**.

The pill for **Remittances per Capita on the Rows** shelf now has a small delta sign to the right of the name, which signifies that it is a table calculation.

10. Change the mark type for **Remittances per Capita** to a bar.

11. You can create more context by adding another reference point, namely the GDP. Drag **GDP (current USD)** to the secondary y axis, as shown in following screenshot, and then drop it when the secondary y axis has a dashed, horizontal line:

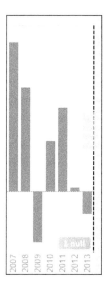

12. Right-click on the y axis for **GDP (current USD)**, click on the **Mary** type, and select **Line**.

13. The following screenshot shows the fluctuation in remittances, as it relates to the global GDP. In 2009, when the global economy was reset, the GDP dropped, and so did the remittances:

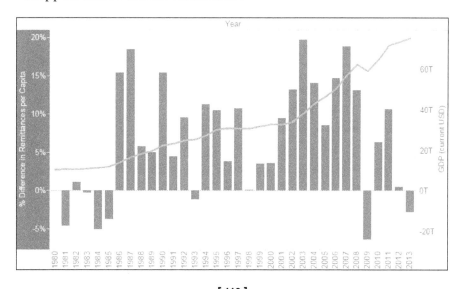

Compute using

The graph from the previous exercise calculates the percentage difference from year to year, which means that the value for each year is being computed relative to the previous year. We can ask Tableau Public to calculate the percentage difference from different years by clicking on the **Context** menu for a table calculation of this type, and then selecting **Relative to**. We can calculate the change from the first, next, previous (the default one), or last values in a partition.

In the following graph, which is a revision of the previous one, the bar height for each year shows the aggregated percentage difference in **Remittances per Capita** since 1980, which is the first year in the visualization. You can do this by performing the following steps:

1. Click on the **Context** menu for the **Remittances per Capita** field on the **Rows** shelf.

2. Click on **Relative to**.

3. Select **First**, as shown in the following screenshot:

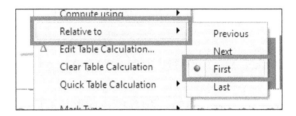

Then, perform the same calculation on **GDP (current USD)**. We changed its use on the **Rows** shelf into a quick table calculation for the percentage difference and set it to **Relative to** for the **First** value in the partition.

We took one extra step—we right-clicked on the secondary y axis, where the **GDP (current USD)** table calculation resides, and selected **Synchronize Axis** so that the axis ranges of both the y axes are the same. This is an important step because we want to make sure that the consumers do not perceive a relationship that is different from the one that we intend to communicate.

In the following graph, you will see that just around the turn of the century, the GDP rate of change since 1980 slowed down, and the rate of change of Remittances per Capita overtook it:

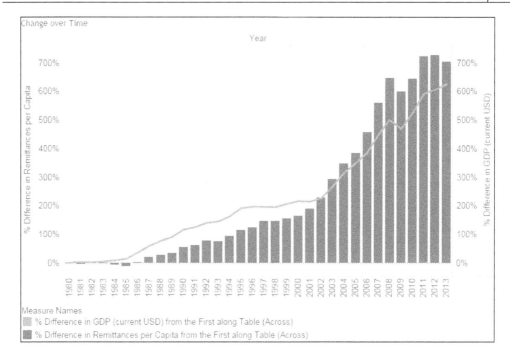

Moving average

The moving average is a powerful table calculation because it gives you the ability to determine how many previous and future values are included in an average.

For instance, we can change the percent difference table calculation that we performed on **Remittances per Capita** by performing the following steps:

1.   Click on the **Context** menu of **Remittances per Capita** on the **Rows** shelf.

2.   Click on **Clear Table Calculation**.

3.   Click on the **Context** menu of **Remittances per Capita** again, select **Quick Table Calculation**, and then choose **Moving Average**.

We cleared the table calculation for **GDP (current USD)** in the following graph as well.

The following graph shows the moving average of Remittances per Capita since 1980 versus the GDP:

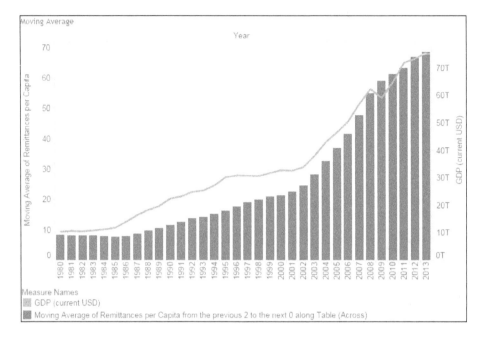

One thing that we don't know is exactly how *smooth* the line is. The term "smooth" means that more than one value is included in the calculation, and if we edit the table calculation, we can control the smoothing. You can also create a parameter that gives users a control over the smoothing. More on this will be discussed later in the book.

Editing table calculations

So far, we have used only the default settings for the table calculations that we have created. Tableau Public gives us several capabilities. Right-clicking on the pill for the moving average of **Remittances per Capita** shows us the following options that are available for this field:

- **Calculation Type**: We can change the type of calculation via this option. Tableau Public automatically creates a different calculated field depending on our selection, and the options available in this user interface also vary based on the selection.

- **Summarize values using**: This gives us the option to select the aggregation type. Ours is a SUM, but we can use AVG, MIN, and MAX in this type of calculation.

- **Moving along** (or **Compute using**, with other table calculation types): This allows us to address the table calculation. In this example, we have only three options, namely **Table (Across)**, **Cell**, or **Date**. The default is to compute by going across the table, which means that Tableau Public computes the values by going from left to right or top to bottom, depending on the location of the dimensions and measures.

- When you click on the drop-down list, you can customize the way the field is computing by telling Tableau Public which fields to **Compute using** the first field, then the second, and so on. In the following screenshot, we have only one dimension, namely **Year**, in use. In case there were others, they would appear in the **Partitioning** pane, and we could add them to the **Addressing** pane and then sort them again. More on this will be discussed later.

- The dimensions that we do not select for **Compute using** are used for partitioning, which is called grouping:

The next feature allows you to tell Tableau Public how many previous and future values to include, and you can also determine whether the current values should be included. Lines that are smoother include more previous and future values as shown in the following screenshot:

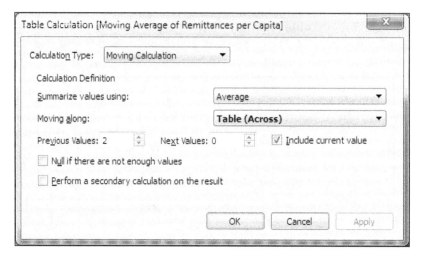

- **Null if there are not enough values**, when selected, does not compute marks for places when there are not enough values. For instance, if you select this for this particular calculation, then the first two years in the visualization will not show a line.

- **Perform a secondary calculation on the result** gives you the power to add additional table calculations. For instance, if this were a running sum that shows the cumulative total, we can use a percentage of the total secondary calculation to show the percentage accumulated for each year instead of the discrete number.

Manually editing table calculations

Learning how to edit table calculations yourself is an advanced capability, but it gives you the opportunity to create rich metrics, such as the percent difference from an average.

In the first example, we will modify the running sum of **Remittances per Capita** so that it includes the preceding and next three values, which can be done from the edit table calculation dialogue by entering those numbers in their respective places.

In the second example, we will modify the **Remittances per Capita** percent difference table calculation to show the percent difference from the average.

Let's begin with the first example. We'll modify the running sum in the following way:

1. In order to edit the table calculation, duplicate the sheet on which we created the original moving average worksheet, which maintains the integrity of the work that we have already done.

2. Then, drag the pill for the moving average of **Remittances per Capita** from the **Rows** shelf to the **Measures** shelf, which prompts us to rename it. Let's call it **Smoother Remittances per Capita**.

3. The original table calculation, as shown in the following screenshot, has one function, one expression, a start offset, and an end offset:

   ° The WINDOW_AVG function tells Tableau Public that the field can be addressed and we want it to apply the AVG function to the values

   ° The expression, in this case, is the field on which the calculation is operating

   ° The start offset, by default, is two places prior to end offset and hence, it has the negative sign

   ° The end offset, by default, is at the current value, as shown in the following screenshot:

4. Modify this calculated field to include three previous and three future values by replacing -2 and 0 with 3 and 3 respectively, as shown in the following screenshot. You can also create an integer parameter and allow users to set the number. We will show you how to do that in the chapter on parameters:

The result is a calculated field that is much smoother than the original one.

In the second example, we'll calculate the percentage difference from the average **Remittances per Capita**. First, we will show you the original calculated field, and then, we'll show the changes that we made.

We duplicated the sheet on which we were working and then dragged the percent differences of the **Remittances per Capita** field to the **Measures** pane, where we renamed it to **Remittances per Capita % Diff from Avg**, as shown in the following screenshot:

In the formula box, Tableau Public calculates the value for each year, subtracts the value of the first year from it, and divides it with the value for the first year.

The following are a few new functions in this formula:

- **ZN**: This means that Tableau Public uses 0 if the value is null.

- **LOOKUP**: This finds a value specified by an offset from the current value. In this case it's -1, which means the previous value, but it could be anything. The other values are the **FIRST()** and **LAST()** functions.

- **ABS**: This takes the absolute value of the previous value. When writing table calculations, you should always use **ZN** and **ABS**, even if you think you don't need them at the time, because you do not necessarily know what will be in your data set in the future.

In order for this to be the percent difference from the average, change the references to the first value to the references to the **WINDOW_AVG**.

- Replace **LOOKUP** with **WINDOW_AVG**.

- Then, delete both instances of the string and the **FIRST()** function, because in this case, the **WINDOW_AVG** does not need an offset, like we had in the previous exercise. We want to use the average for the whole partition.

The new formula looks like the one shown in the following screenshot:

Ranking

Our examples so far have focused on comparing measures that occur along a time continuum. Table calculations are useful for a comparison between different dimensions as well. Ranking by discrete numbers, as well as percentiles, is powerful. Many people want to know the top or bottom number of members in a dimension. In the next example, the final product shows the bottom 20 percent of the countries in each region according to their availability of improved water sources, and each country will be colored by its percentile of life expectancy.

Start off by creating a basic bar graph from the World Development Indicators data source that shows the maximum percentage of a country's population that has access to improved water sources. Drag **Region** and **Country** to the **Rows** shelf. Then, when we drag **Improved Water Source** (%) to the **Columns** shelf, we aggregate it as a maximum rather than a sum. Select **MAX** based on the assumption that unless there's a major natural disaster (such as floods and earthquakes, which were used as data points in the previous chapter) or war, it's unlikely that the access to clean water will decrease.

Then, sort the countries in the ascending order by clicking on the sort icon on the *x* axis.

The real ranking tasks begin here, with the label. The objective is to show the bottom 20 percent by region. So, we need to do several things along the way that help explain the ranking functionality. Perform the following steps:

1. Drag **Improved Water Source** (%) to the **Label** shelf.

2. Then, click on it and select **Quick Table Calculation**.

3. The Quick Table Calculation that we selected is **Rank**. Then, drag the field to the **Measures** pane and rename it to **Country Rank – Water**.

4. Check out the snippet of the visualization in the following screenshot; each country is ranked in the descending order by the percentage of citizens who have access to clean drinking water:

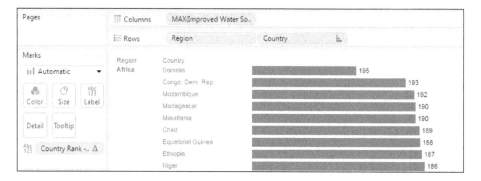

5. Note that the countries are ranked as a part of the whole table. We would like to rank them by region. Therefore, click on the pill, which is present on the **Label** shelf, select **Compute using**, and choose **Pane** (down).

6. To the lower-right of the visualization, there is an indicator that we have several null values. This is because some of the countries have not reported these data points. Click on the indicator to exclude the countries with null values.

7. There are other functions, such as INDEX(), that perform similar tasks, except that INDEX() actually produces the row number in the partition and not the rank. If you change the sort order of the countries shown, the labels stay the same; wherever you see Somalia, it will be ranked 53 in Africa.

The next task is to show the bottom 20 percent of the countries. Showing the top 20 percent would not add much value since the measure cannot exceed 100 percent of the population. In order to do this, we created a new calculated field called **Percentile Filter**, and told Tableau that we want to know whether the percentile of each country is less than 2. This means that it's in the bottom 20 percent of all the countries in its partition. Then, define the partition. The default is that the entire table is the partition, but we want each pane to be the partition.

The filter shown in the following screenshot has three possible results, namely true, false, and null, which makes it a Boolean field that shows up with a **T | F** icon in the data window. Since we excluded the nulls, we will not get null values:

In order to make sure that the total ranking of the countries in each region and showing only the bottom 20 percent, perform the following steps:

1. Click on **Default Table Calculation**.
2. Select **Country**.
3. Click on **OK**.
4. Click on **OK** again.
5. Drag the field to the **Filters** shelf.
6. Select **TRUE**.

You can also allow users to see the percentage of their preference by creating a parameter with a data type of float and a rank from 0 to 1.00 that replaces the hard-coded .2 in the formula, as shown in previous screenshot.

There are several functions within Tableau Public for ranking, and except for RANK_PERCENTILE, the other functions vary only in the way they use unique or duplicate ranking values.

Window versus running functions

We have briefly discussed the WINDOW functions already. There's also another type of function, called the RUNNING function. The WINDOW functions have specific partitions, that is, they measure either the whole table, pane, or cell, or from a specific number of previous or future values. Alternatively, the RUNNING function compares all the values before the current value in the partition.

In the following example, we graphed the average percentage of the GDP that was from high-tech exports by region and by year in the World Development Indicators data source. This is a simple spark bar graph. We have hidden the header for the y axis in this graph.

We created a calculated field that, for each year, tests whether the average **High Tech Exports** as a percentage of the GDP is equal to the running maximum, that is, is the value for any year higher than all the years before it. The results are true and false, as shown in the following screenshot:

The RUNNING_MAX function, which is one of several RUNNING functions, does not have the start and end offsets like the WINDOW functions. However, you still need to address it.

A note on addressing

The following table calculation computes across the pane. You can also set it to **Compute using Year**. Year is the field along the x axis. Addressing the table calculation across the pane tells it to compute across the x axis and then to use other fields on the visualization, such as **Region**, as the partition or group. So, this field is addressed across year, and the calculation is performed for each **Region**.

Add the **Region** field to the **Color** shelf in the visualization and then format the true values so that they are blue, and format the false values so that they are orange, as shown in the following screenshot:

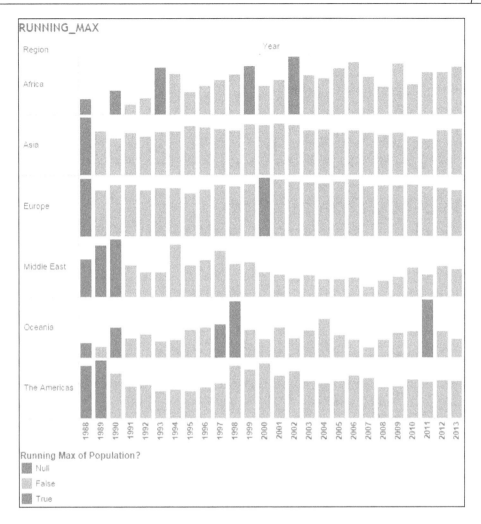

The level of detail calculations

The **level of detail calculations** (LOD) enable you to tell Tableau Public at exactly which dimensional level you would like to aggregate without having to place that dimension on the visualization. Additionally, you can include or exclude dimensions in the calculations and create calculations that include all the underlying data.

The following are the three types of LOD calculations:

- **FIXED**: This computes a value for the dimension that you specify
- **INCLUDE**: This computes at a dimensional level and is not included in the visualization itself
- **EXCLUDE**: This excludes a dimension that is a part of the visualization

LOD has a big caveat. While you can perform table calculations on them and create aggregations and functions within them, you cannot include table calculations in them.

A FIXED LOD calculation

Each LOD calculation has the following three features that are different from those that we have discussed before:

- The **LOD** expression type: In this case, the LOD expression type is **FIXED**
- The **Dimension** on which the calculation is operating (you can add dimension levels, which should be arranged in the increasing order of granularity by separating them with a comma): In this case, it is **Region**
- The **Aggregation**: In this case, it is **COUNTD([Country Name])**

Note that the entire expression is enclosed in curly braces, as shown in the following screenshot. The LOD calculations are the only instances in Tableau Public where you will use curly braces:

This field provides us with the unique number of countries in each region, even though 80 percent of these countries are filtered out. This provides great context.

In the following example, we will discuss:

1. Add this field, which is a measure, to the **Rows** shelf to the right of **Country**.
2. Then, click on its **Context** menu and select **Discrete** instead of **continuous** so that it becomes a dimension.

3. Drag the field so that it's to the right of **Region** and to the left of **Country**, as shown in the following screenshot:

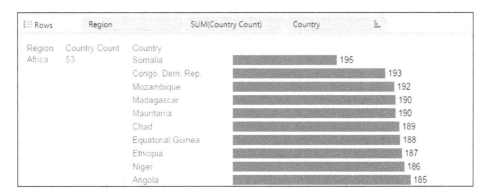

This visualization shows us the bottom 20 percent of the 53 countries in Africa in the data set. It would be even more helpful to know what the overall access to improved water sources is in each region and not just by country.

The following are a few issues that we need to overcome with virtualization:

- The data is not necessarily set up for this.

- The granularity of the data is at the country level, and we only know the percentage in each country that has access to clean water. We don't know the discrete number.

- We do not have the totals by region. We need to determine the overall number of people with access to clean drinking water, and then we need to divide this number by the population of each region.

The INCLUDE and nested LOD calculations

In the following example, we will show you how to solve some problem by using an INCLUDE LOD calculation nested within a FIXED LOD calculation, and each will be aggregated. This is an issue that most people have had to solve by using data sources with different levels of aggregation, which takes a lot of time to develop and render in Tableau Public.

Create a new field within the World Development Indicators data source and name it **Population with Access to Improved Water**. It's kind of a long name, but we want to ensure that no one misunderstands the metric.

So, here's our approach to calculating the overall percentage of each region with access to clean water:

- The primary LOD type is **FIXED** at the **Region** level.
- The numerator (enclosed in the red box shown in the following screenshot) adds up the total population that has access to clean water.
- The secondary LOD type is INCLUDE at the **Country Name** level. We are using INCLUDE because we want to roll up the country-level data to the region.

Rationale: For every country, we need to know the total population that has access to the sources of clean water. We have the percentage of the population. So, we need to multiply the maximum percentage with the total population to get the number of people who have access to clean water.

- The denominator (enclosed in the blue box in the following screenshot) is the total population. We need to know the total population by region. So, we add up the total population of each country.
- We will use the INCLUDE function at the **Country Name** level and then use the MAX aggregations. The granularity of the data is at the **Year** level. We are not including Year, because that would produce the population at the **Country Name and Year** level. We just want the total population of each country, as shown in the following screenshot:

```
'opulation with access to improved water      COmplete WDI                                    ⊗

{ FIXED Region:
  sum(

  { INCLUDE [Country Name]:

  max([Improved Water Sources (%)])*max([Total Population])}
  )
  /
  sum(

  { INCLUDE [Country Name]: max([Total Population])}
  )
}

The calculation is valid.                                    Apply        OK
```

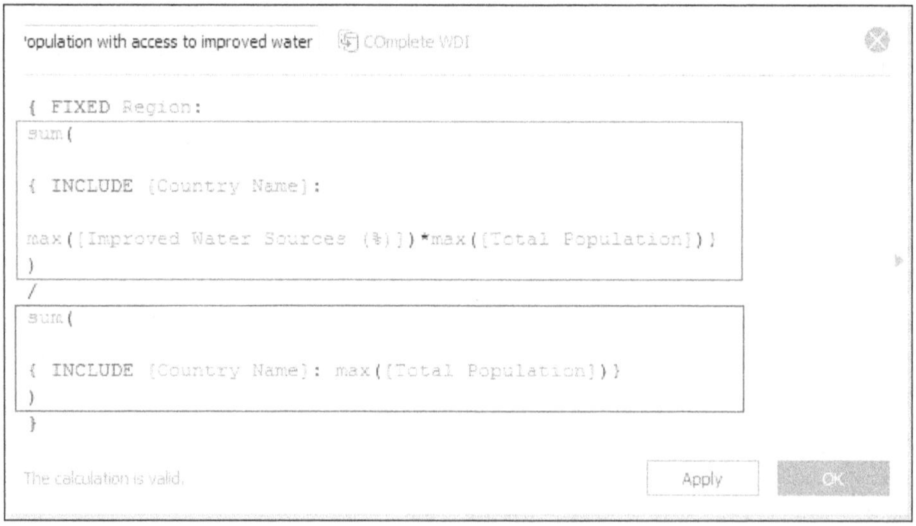

This is a long field, but it accomplishes what has been difficult before, namely aggregating at multiple dimensional levels in one calculation without duplications, and aggregating records that are not a part of the visualization.

Add this field to the **Rows** shelf of the previous bar chart, turn it into a discrete field, and drag it to the left of **Country Name**, just as we did for **Country Count**.

Also, right-click on the new calculated field, click on **Default Properties**, and then select **Format** so that we can set the number format as a percentage.

Here's the finished product. Only 69 percent of the people in Africa have access to clean drinking water:

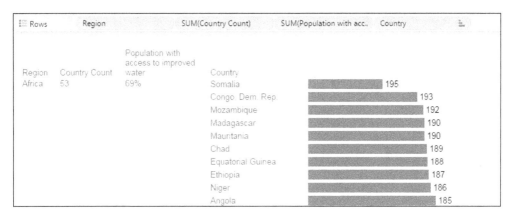

Summary

In this chapter, you learned how to create quick table calculations to add depth and context to your visualizations. You also learned how to edit table calculations, how the different types of table calculations function, and how to address table calculations and create them manually. One of the important features of data analysis is the comparison of a data point to the others around it so that users understand the context of its performance. You learned how to create a basic, and advanced level of detail calculations to aggregate data points that are not visible, and to aggregate at multiple dimensional levels as well.

In the next chapters, which focus on mapping and parameters, we will discuss using many of the concepts that you learned here to give control to your users and illustrate compelling, rich stories with contextually curated visualizations.

7
Dashboard Design and Styling

The purpose of a dashboard is to tell a story. In the scheme of human history, hieroglyphs, the Bible, epic poems, theatre, comic books, and even social media are a few of the many different media through which we recount events to each other. All stories have one of the several primary purposes, to teach values or historical lessons, create culture, motivate people to take action, or entertain, to name a few.

Knowing the purpose of the story that you are telling is imperative to making it memorable. Good stories have several important attributes, which should be kept in mind when you are working. Some of these attributes are as follows:

- They're simple and focused
- They create an emotional response
- Listeners can relate to the actors in them
- They have good, real data

It's a good idea to take a few minutes to think about the story that you want to tell before you even create a new dashboard. Think about the stories that are the most memorable to you and the data points in them. Then, put yourself in the place of your listener. Design with your listener in mind. If you can develop a list of the top three data questions that you think are most important to your listener and then design a dashboard that focuses on answering them simply and evocatively, then your dashboard will likely be impactful.

With Tableau Public, you can tell stories in a variety of ways—with individual visualizations, well-formed dashboards, and story points. You can also integrate your work in Tableau Public into other tools by using the JavaScript API (which is beyond the scope of this book). But just because there are many tools and options at your disposal, it does not mean that you should make your dashboards overly complicated. It's actually just the opposite. Good journalism focuses on keeping stories simple, and you should use the same ethos with your stories. *Just because you can do something doesn't mean that you should do it.*

It truly applies well to designing dashboards; less is more. Knowing your user and their questions is fundamental to a good design, and everything else flows from that.

In Tableau Public, you can make a dashboard from one or more visualizations, and you can add images, free text, filters, legends, parameters, links, and even embedded web pages. All of these capabilities are designed to enable you to create a data story that has the right context for your dashboard, both for the consumer and for its relativity to other events and data elements. While you can publish individual worksheets to Tableau Public, we encourage you to organize your work into dashboards so that you can guide your user through a complete story.

In this chapter, we will discuss some best practices for dashboard creation, the elements of a dashboard, and how to navigate the dashboard worksheet. The following topics will be discussed in this chapter:

- The design process
- Best practices of dashboard design
- The elements of a successful dashboard
- Creating a dashboard
- Adding context with titles, images, links, and tooltips

In the next chapter, we will discuss dashboard filters, actions, and parameters, and after that, we will discuss publishing your work to Tableau Public.

The dashboard design process

The purpose of Tableau Public is to help you tell data stories interactively. Dashboard design, like view design, should tell a concise story that flows from one element to the next. When you are first starting your journey of dashboard design in Tableau Public, you should expect it to be an iterative process in which you design dashboards and revise them over time, often with input from other people. You might also find it helpful to draw out dashboards before building them in Tableau Public. Leave room and time for experimentation when building your dashboard.

You may find that more views, or different views, are needed in order to tell your story. Pulling various views together into a dashboard sometimes exposes flaws in the initial design. In that case, go back and change or add to your views, and then come back to the dashboard design process.

Best practices for dashboard design

Dashboard design should highlight data and decoration should be limited, to add to a reader's understanding. Much has been written about dashboard design for data visualization. The following are some key points that you should consider when designing dashboards:

- **Keep it simple**: Your dashboard needs to answer only one question. If you can construct it so elegantly that it answers several questions, that's great, but according to *Stephen Few*, the maximum number of data points that any of us can remember at once is three.

- **Keep in mind the audience**: Consider what devices will be used to read or consume the data story and content.

- **The method of consumption**: Will users interact with virtualization on their computer, mobile phone, or tablet? Size your dashboard accordingly.

- **Performance**: Like any other web application, the amount of time it takes for a consumer to get what they need is a major factor of performance. Design your visualizations and data sources so that they perform efficiently.

- **Start big and end small**: Always place the most aggregated (macro) data points or summary metrics on the upper left, and guide users to granular, actionable data points.

- Use **colors**, **graphics**, and **fonts** that are appropriate for your subject matter and consumers. Choose a simple, single color or complementary palette colors. Consider red-green color blindness and improve accessibility. Remove all non-data ink except labels, limited titles, and bare-bones instructional text. Additionally, you can create your own custom color palettes in Tableau Public fairly easily. Use labels, titles, and simple instructional text for the dashboard, and format tooltips so that they add context and calls to action for your users.

For the dashboard that we are using in this chapter, we created a rough wireframe in Microsoft Visio. We'll refer to it throughout this chapter so that you can see how the functional elements that we are discussing will be included in the final design. It's fairly common for people to draw out the designs that they would like to build in Tableau Public. Also, it's a great way to keep your objectives in perspective when laying out the dashboard in Tableau Public.

Our sample dashboard, which will be included in a blog post at `www.dataviz.ninja`, follows the **Rule of Thirds** — after adding the title to the dashboard, the three primary visualizations, namely aggregated data, change over time, and granular details, will consume about one-third of the total space, as shown in the following screenshot. We have also provided space for filters and color legends, and we will continue to use this dashboard design concept in the following two chapters, which cover filtering and publishing:

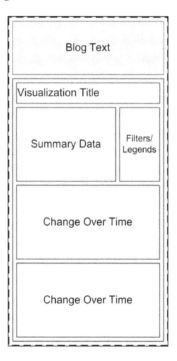

Creating a dashboard

Creating a Tableau Public dashboard is a process of adding the various elements (starting with worksheets), onto the dashboard canvas, adding titles, configuring filters, titles, and layout, and then modifying the behavior of the dashboard.

Now, we will build a dashboard from simple visualizations that were created by using the Climate Change data that was published by the World Bank and which is available at `http://data.worldbank.org/data-catalog/climate-change`. It's structured in a way that is similar to that of the other data sources that we are using. You can download the sample dashboards from `https://public.tableau.com/views/Chaper7-Dashboards/CO2EmissionsDashboard`. We will use the data sources that we have discussed in the previous chapters to build sample dashboards so that you can download the source files to practice.

Our workbook includes a summary map of carbon dioxide emissions by country in 2008, as well as a line graph that shows change over time as regards the total emissions and emissions per capita by region. Lastly, it contains a heat map that shows the foreign aid distributed for generic programs by the United States by year.

The objective of this section is to build a dashboard that allows users to navigate from high-level areas of interest, that is, outliers, to more detailed data to identify causation.

To start the dashboard creation process, click on the dashboard tab at the bottom of the Tableau Public interface. The dashboard tab is the tab with an icon of a rectangle divided into four quadrants, which is outlined in the following screenshot.

You can also browse **Dashboard | New Dashboard** from the **Menu** bar:

When you select the dashboard tab, the dashboard view is displayed. This is the interface that allows you to build and configure dashboards from the worksheets that you have already created in this workbook. To the left of the window are the available worksheets. From the left side of the window, you can also add to the dashboard various supporting elements such as text boxes, labels, web page windows, and images. To the right is the canvas area of the dashboard, where sheets and objects are dragged to, arranged, and configured.

The dashboard tab interface

The dashboard interface is similar to the worksheet workspace in many ways. In the following screenshot, we have identified its several key areas.

Like the worksheet view, the dashboard has the following elements:

- The menus and toolbar, which are placed horizontally across the top of the page
- The **Dashboard** pane on the left, where you can determine the exact composition of your dashboard
- The canvas, where you will drag worksheets and other visual elements

We have numbered the elements on the dashboard, as shown in the following screenshot. They are as follows:

- A list of worksheets **(1)** that have not been hidden and which are sorted according to the order in which they are arranged in your workbook
- Containers and objects **(2)** that can be dragged onto your dashboard to increase the integrity of the design
- A controller **(3)** for new objects
- A hierarchy of dashboard objects **(4)** that can be used for navigation
- A controller **(5)** for the dashboard size
- The canvas **(6)** where you will compose the dashboard

If you are creating a new dashboard, the canvas will be blank. If you open an existing dashboard, the various existing dashboard elements will be displayed on the canvas. The following screenshot shows the dashboard interfaces:

It's important to keep in mind the purpose of your dashboard. We are building a dashboard that tells the story of carbon dioxide emissions over time. The dashboard should allow users to navigate to the areas of interest and then understand the factors of consumption.

The dashboard will have the following key elements, which were previously discussed:

- Title, so that our users know what they are viewing
- At least one visualization with data and proper context is required so that different data points relate to each other and to the user
- A call to action (conclusion) is also needed so that users know how to explore the details within the dashboard or go to another web page

In order to build this dashboard, we will start with a single visualization, which is the visualization with summary data in this case. Drag a sheet onto the canvas by hovering over the visualization that you want to add, pressing the left mouse button, and dragging it onto the canvas, as depicted in the following screenshot. You can also double-click on the worksheet name.

If you wish to remove a worksheet from the dashboard, click on the title of the visualization until a grey container bar appears at the top border of the visualization. Then, right-click on the grey bar. Select **Remove from Dashboard** from the shortcut menu:

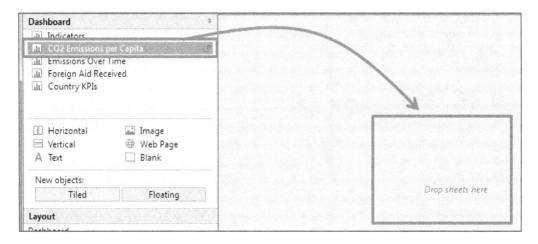

The result, as shown in the following screenshot, shows the map that we created. This map shows the CO_2 emissions per capita in 2008, which was taken from the World Bank official website. Each country is colored by its relative percentile in the data set, and the label shows the CO_2 emissions per capita.

However, users don't know all of that yet:

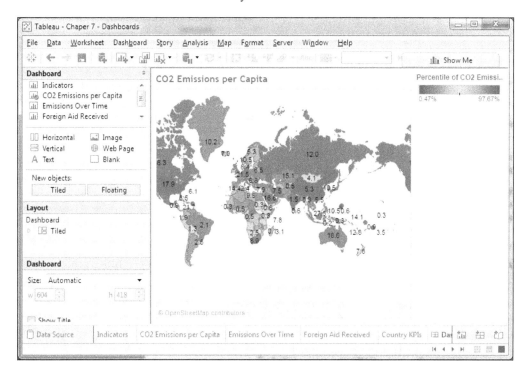

Tableau Public automatically added the following two elements:

- The worksheet name as the title
- The color legend in a vertical layout container to the right

In order to compose an empathetically designed user experience with proper context, we need to use layout containers to rearrange the dashboard and create a flow of both navigation and information, that is, a data story. The next section discusses layout objects, and we will show examples of how to use them.

Layout objects

Tableau Public has several different layout objects that you can add to your dashboard to control the composition.

Objects are versatile, and you can use them to reflect the overall theme of your graphic composition. From the context menu of an object, you can perform the following tasks:

- Format it with borders and background colors
- Remove it
- Set it as a floating or tiled object
- Deselect it
- Remove it from the dashboard

You can set the height and width attributes for floating objects, just like you can for worksheets that you have added (and for the dashboard as a whole, which is highly encouraged).

The following objects can be added to your dashboard:

- Horizontal layout containers: You can drag worksheets that you want next to each other into these containers.
- Vertical layout containers: You can drag other objects and worksheets that you want to stack from top to bottom into these containers.
- Text objects: These can be used to add titles and calls to action. Though you can't add field tokens to text objects, you can add parameter tokens, which will be discussed in depth in the next chapter.
- Images: You can use these to browse logos or branding elements that add richness to your dashboard. You can also use images as links to web pages, since each image can have a URL attribute.
- Web page objects: These can be used to add content from the Internet. You can even create dynamic content, which will be covered in the next chapter too.
- Blank objects: These objects can be used to control space. Blank containers are transparent. So, the background colors that you are using will show through them.

In the section of the dashboard pane that says **New objects**, the default object is set to **Tiled**. You should leave it there until you are comfortable with dashboard design. Each new object that you add to your dashboard will have this setting, and though you can easily change the attribute for individual sheets on a dashboard, it's best to keep things simple in the beginning. The **Tiled** automatically aligns objects to the dashboard grid, and floating adds new objects to the dashboard that are detached from the dashboard grid.

When you add objects, like worksheets, into a horizontal or vertical layout container, the widths and heights of the objects will be sized automatically, unless you specify otherwise. In case you're using actions, which will be discussed in the next chapter, it's particularly useful to use containers for sizing automatically.

Setting the size of dashboard elements

Usually, dashboard elements such as containers can be set by manipulating the border handles of each container and resizing them to a suitable size. Tableau Public also offers a **Size** feature, as shown in the following screenshot, that allows for more control in the positioning and sizing of these dashboard elements:

The dashboard element must be in the floating mode so that it can be sized and positioned. In case the container or element is not already in the floating mode, it can be set to floating by clicking on the dashboard element and checking off the **Floating** checkbox, as shown in the previous screenshot.

To change the positioning of the dashboard element, enter the **x** and **y** coordinates (this will take some trial and error) in the corresponding **Pos:** fields, or scroll up and down with the arrows for fine correction. To change **Size**, enter the width (**w**) and height (**h**) values in the corresponding fields using the up and down arrows for fine adjustment.

In the **Size** section of the Tableau Public controls on the left side of the dashboard, you can also choose to show the title of a particular chart by selecting or deselecting the **Show Title** checkbox. You must first select the dashboard element either in the dashboard, or in the **Layout** section of the Tableau Public controls. To show or hide the title of the dashboard, you can click on the **Dashboard** field in the **Layout** section and select or deselect **Show Title**. This may also be done, as described earlier in this chapter, by using the **Dashboard | Show Title** menu command.

Sizing the dashboard

Tableau Public presents many size options for the entire dashboard. This feature helps you select a size either for a blog or other website, or optimized for a tablet computer. To set the dashboard size, click on **Dashboard** in the **Layout** section of the Tableau Public controls.

There is also an option to resize the entire dashboard. For example, you may find that scrollbars appear to the right, or at the bottom of the dashboard, that you don't want to see. Alternatively, you may prefer matching the dashboard to a standard web page or computer desktop size. Again, in the **Layout** section, select **Dashboard**, or ensure that no worksheet or other elements are selected on the dashboard itself. The **Size** section presents a drop-down menu with a variety of dashboard size options.

Many sizing options are available for selection depending on your needs. **Blog**, **iPad**, **Laptop**, **Desktop**, **Exactly**, and **Automatic** are some of the typical settings. It will take some trial and error to properly size a dashboard for your needs.

It is important to consider the methods in which you will be publishing your dashboard. If you are publishing your dashboard to your blog, you need to make sure that it properly fits into the layout that you have selected.

Building a dashboard

In order to build a sample dashboard, which has a title, aggregate data, change over time, detailed data, and proper context, as shown at `https://public.tableau.com/views/Chaper7-Dashboards/CO2EmissionsDashboard`.

Perform the following steps to build the dashboard:

1. Create a new dashboard by clicking on the dashboard creation icon in the filmstrip.

2. Name it **CO2 Emissions Dashboard**; the name is important because it will be part of the URL for the workbook.

3. Size the dashboard so that it fits a blog, as shown in the following screenshot. This size fits in a typical WordPress blog, and it is also narrower than most mobile devices and laptops. Therefore, the size will work well for users of multiple devices:

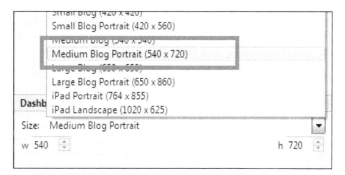

4. We added containers knowing that we wanted to have a title, aggregate data, supporting details, and conclusions. We added a vertical layout container so that our visualizations and objects will stack on top of each other. The following steps will guide you to add data-map in the layout container:

   1. We dragged the aggregated data-map of **CO2 Emissions per Capita** into the vertical layout container.

   2. We then dragged the next worksheet, **Emissions Over Time**, into the very bottom of the vertical layout container so that the worksheet and the map would be automatically sized.

   3. We dragged the worksheet from the **Dashboard** pane to the bottom of the container and dropped it when we saw the wide gray border, as shown in the following screenshot:

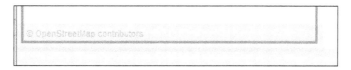

5. We assessed that the dashboard shown in the following screenshot isn't pretty. It still needs a title, and it should be rearranged:

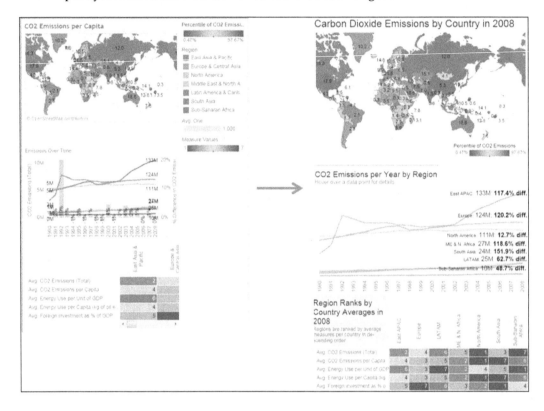

6. We added a title before rearranging it, by performing the following steps:

   1. Drag a Text object to the very top of the vertical layout container above the map.

   2. Populate the title (**Carbon Dioxide Emissions by Country in 2008**) and then format it in 18 pt black Arial, which is wide enough to fit in the screen.

7. The layout needs some work. Color legends need to move closer to their respective visualizations, and we need to change it so it can be fitted in some visualizations, as shown in the following screenshot:

8. Move the **Percentile** color legend for the map by performing the following steps:

   1. Click on it to select it.
   2. Click on its context menu.
   3. Change it to floating.

9. Move it below the center of the map (we will move it again later).

10. We decided to remove the color legend for **Region** altogether.

11. Remove the size legend for **Avg**. Once we have actually used a bar in the heat map; you can check it out by clicking on it.

12. We removed the last color legend, which is for the heat map. The colors are used to indicate rank, and they follow the same color scheme as that of the map (brown is bad, and blue is good).

13. Adjust the fit of the containers as:

   ° The heat map needs to be fit differently. Therefore, click on its context menu, select **Fit**, and choose **Fit Width**.

14. Next, add titles by performing the following steps:

15. Add a text object above the line graph; the title should be highly descriptive. It tells our users exactly what they are looking at, and it also tells them how to get more details. We used 14 pt black Arial font for the main title, and the call to action (that is, the text that tells the user what to do) is in 10 pt gray Arial, as shown in the following screenshot:

 You can also add an image above the line graph. We added an image of a simple gray line that we created in Microsoft Paint and then saved as a bitmap.

   ° We also added a floating title for the heat map because we had extra horizontal space but less vertical space. We used the same fonts as that of the previous title and added a light gray shade so that it's visually separated from the graph itself.

16. Add more context with tooltips and labels; it's important for users to know exactly what they are looking at, how the data points relate to each other, and how it relates to them. You can modify the tooltip for the map by performing the following steps:

   1. Click on **Map Visualization** to select it.

   2. Click on the **Worksheet** menu.

   3. Select **Tooltip**.

   4. Modify the contents and font so that they add obvious context and do not require users to do a lot of work to figure out what they are viewing, as shown in the following screenshot:

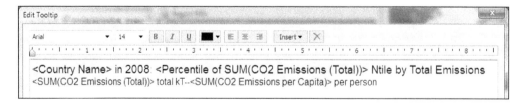

5.   Repeat these steps for other visualizations. Make sure that you use the same font face and size in each tooltip. Consistency of design is critical.

6.   Modify the data labels, particularly for the line graph.

7.   Navigate to the worksheet, and place the appropriate fields on the Label shelf in case they were not there already.

17. Format dashboard so that the right emphasis is created.

Change their location relative to each other on the dashboard so that they are all legible by clicking on individual labels and then dragging them elsewhere.

We felt comfortable making custom placements because we do not have a filter on the dashboard. (We will discuss filters in the next chapter.) We are confident that, with a specific size of the dashboard, the labels will always appear where we placed them.

If you want to undo custom placements, you can right-click on the label and select **undo the custom location**, as shown in the following screenshot:

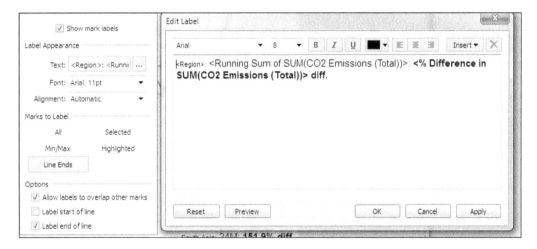

○   Adjust the size of the dashboard to 540 x 900 pixels because it appears compressed. Unfortunately, this means that we'll have a vertical scroll bar when not using a studio monitor, but we needed to make sure that the visualizations were legible. It's better to have a vertical scroll bar than a cramped dashboard. If no one can read it, it's a waste of time and effort.

Before publishing the dashboard, right-click on every worksheet in the filmstrip and select **Hide**.

This dashboard, as available at `https://public.tableau.com/views/` `Chaper7-Dashboards/CO2EmissionsDashboard?:embed=y&:display_` `count=yes&:showTabs=y`. It tells the story of not only how carbon dioxide emissions have changed geographically over time, but also what compounding factors may be contribute towards the future growth of Tableau Public.

Summary

In this chapter, we discussed best practices of dashboard design as well as how to build an empathetically designed user experience that tells a concise story with data. You also learned that Tableau Public dashboards are composed of one or more sheets (chart visualizations) with added elements such as text, captions, and interactivity such as filters and actions. Dashboards may also contain, or open web pages and can pass dashboard information into the web page in some instances.

Keep in mind that a good dashboard design helps you convey a pertinent message to the reader. Have a goal in mind pertaining to what you wish to get across to the readers. Elements of a good dashboard design start with good data and visualizations, and these visualizations come together in an aesthetically pleasing, intuitive dashboard that helps users tease out conclusions and discoveries of their own. Knowing your audience is critical to dashboard design.

In the next chapter, we will explore filters and actions, which are important tools in adding interactivity to dashboards.

8
Filters and Actions

Dashboard filters and actions in Tableau Public enable users to precisely select the information that they would like to explore, which is powerful. Selecting what's interesting makes a data story personal and relevant to users; all of this is done without shifting their attention or their mouse away from the dashboard. Filters and actions create true interactivity. They also give you, as an author, the ability to connect disparate sources of data or web pages to complete user experience.

In this chapter, we will discuss the following topics:

- Adding and using **Quick Filters**
- Filtering the **Data** sources with parameters
- Filtering the **Data** sources with controller worksheets
- Highlighting actions
- Action **Filters**
- URL Actions

Adding and using Filters

Filters are designed to meet several needs of your users; the following are some of these needs:

- Limiting the scope of an analysis
- Allowing users to view only what interests them
- Removing outliers

Using a filter to limit the data that is extracted into your workbook improves the performance of the workbook. However, the following are the two disadvantages of using filters on dashboards:

- Using many filter actions with multiple Dimension members can slow down a dashboard's performance.
- Filters can be applied to only one data source at a time. So, in case multiple data sources are being used on one dashboard, you'll need to use parameters or controller worksheets to filter all of them.

In this chapter, we will use the dashboard that we created in the previous chapter. You can download the finished product by visiting `https://public.tableau.com/views/Chaper7-Dashboards/CO2EmissionsDashboard?:embed=y&:display_count=yes&:showTabs=y`. The dashboard uses only one data source at this point. As we progress through this chapter, we'll add another dashboard along with a fourth visualization.

Adding Filters to worksheets

In the dashboard that we are using, we have already filtered the map to show only the data for 2008, which is a blue pill on the **Filters** shelf, because it is a discrete number. You will also see that there are green pills for **Latitude** and **Longitude**. Both of these pills are continuous numbers, and they are included on the **Filters** shelf because we filtered out the null values, as shown in the following screenshot:

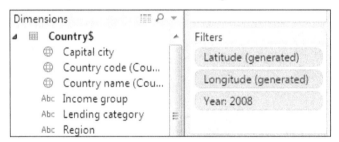

If a field is not already on the **Filters** shelf, you can drag it there from the **Data** window. Alternatively, you can create a **Quick Filter** by right-clicking on it in the **Data** window and selecting **Show Quick Filter**, as shown in the following screenshot. Tableau Public will then display the **Quick Filter** on the upper right-hand side of the workspace, but you can easily move it around the worksheet, as shown in the following screenshot:

For instance, if a field is a part of a visualization and it's on the **Rows** or **Columns** shelf, the **Marks** card, or anywhere else, you can click on its **Context** menu and select **Filter**.

There is a catch to adding filters. If a field is not on the **Filters** shelf of a worksheet, you won't be able to add it as a filter on the dashboard. In the following screenshot, we navigated to the worksheet in the workbook that has the map on it and added the filter there. In the next section of this chapter, we will go back to the dashboard and add the **Quick Filter**.

To add a **Quick Filter** for **Region**, perform the following steps:

1. Right-click on **Region** in the **Data** window.

2. Click on **Show Quick Filter**.

3. Then, it's time to make a selection. With a discrete field such as **Region**, you will see the dimension members that you can choose. The **Filters** dialog box is shown in the following screenshot:

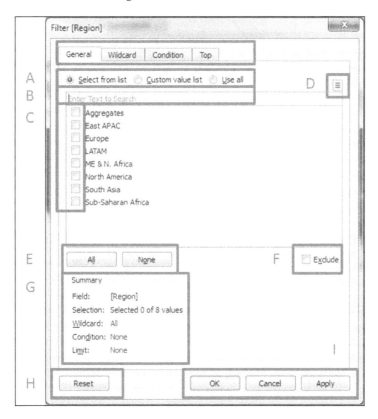

4. In the **General** tab, we can perform the following operations, as shown in the preceding screenshot:

    °   **Select from list** (**A**): This uses either a custom list of values, or all the values.

    °   The search box (**B**): We can search for a text string that is present in a dimension member.

    °   Checkboxes (**C**): Click on the checkboxes next to the dimensions that we want to keep or exclude.

    °   The filter (**D**): This sets the filter to show fewer values, which means that it shows the values that are automatically limited by the selections of other filters, and it doesn't show all the discrete values in the field.

    °   **All** or **None** (**E**): You can select **All** or **None**, as required.

    °   **Exclude** (**F**): This excludes the values that we have selected. If you are excluding values, be careful when exposing the filter to users. They might need some extra instructions about what the exclusion means and how to use it.

    °   **Summary** (**G**): This shows a summary of your selection.

    °   **Reset** (**H**): This resets all values.

    °   **OK**, **Cancel**, and **Apply** (**I**): Click on **Apply** and then on the **OK** button to filter and set worksheet.

5. In our example, we selected all the values and then deselected **Aggregates**.

6. Then, we clicked on **OK**.

The pill for **Region** (which is blue because it's a discrete field) is now in the **Filters** shelf. When you go back to the dashboard, you can add it to the worksheet.

The other **Filter** tabs allow you to further specify the members that you want to retain or exclude in the analysis. You can perform the following operations:

1. Establish the conditions that need to be met

2. Use formulas to determine the members that you want to include

3. Include the top or bottom members by a certain metric

This works particularly well in graphs other than maps because users are often only concerned with the best and worst performers.

Adding Quick Filters to a dashboard

If your worksheet is already on the dashboard, the new filter will not appear there automatically. The dashboard is laid out in a vertical column for the most part, and we do not have filters on it. Its top looks like the following screenshot:

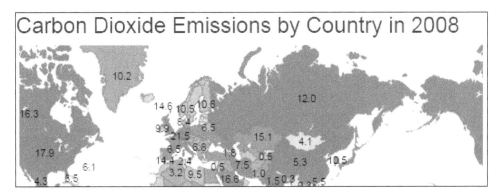

We want to add the **Region Quick Filter** to the dashboard and then apply it to the other worksheets on the dashboard. When we add a filter (or a legend) to a dashboard that we have arranged very specifically, Tableau Public will automatically add a vertical layout container on the far right. Don't be alarmed by this. You can move filters and legends to the locations you want them to be in.

If a field is already a part of a visualization that is on a dashboard because it is one of the fields that is being displayed on the **Detail** or **Tooltip** shelves, you can add it to the dashboard by clicking on the **Context** menu for the visualization and then selecting it from the list of the **Quick Filter** options.

In the next few exercises, we will discuss the following topics:

- Adding the **Region Quick Filter** to the dashboard
- Formatting it to be a drop-down list
- Setting it to show only the relevant values
- Applying it to all the worksheets on the dashboard
- Moving it so that it's in the same container as that of the title of the dashboard

In order to add and modify the **Quick Filter,** perform the following steps:

1. Select the map worksheet on the dashboard by either clicking on the ocean in the map, or selecting it from the list of worksheets in the **Dashboard** menu to the left of the dashboard.

2. Click on the **Context** menu.

3. Hover the pointer of the mouse over the **Quick Filters** and click on **Region,** as shown in the following screenshot:

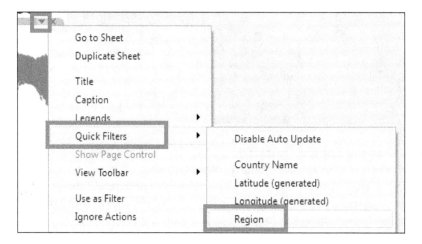

4. The **Region** filter will appear in a **Vertical Layout Container** to the far right of the dashboard. We'll move it later. It's important to be mindful of unused white space.

5. Now, we will change the format of the quick filter. There are many options, and we want the most compact one, which is a multi-select drop-down list. Click on the **Context** menu of the filter and change it to show **Multiple Values (Dropdown).**

6. There are many other modifications that we can make by clicking on the **Context** menu of the filter. Click on the **Context** menu again.

7. Click on **Only Relevant Values** so that it shows only the **Regions** that meet the conditions of the other filters.

8. Apply it to the other worksheets on the dashboard that are using the same data source. From the **Context** menu, click on **Apply to Worksheets.**

9. In the **Apply Filter to the Worksheets [Region]** dialog box, click on **All on Dashboard,** as shown in the following screenshot, and then click on **OK.** Keep in mind that this only applies **Quick Filters** to worksheets using the same data source:

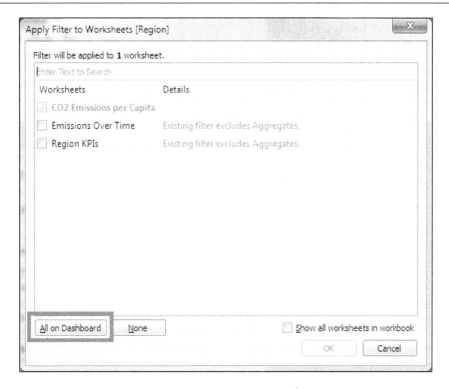

The other modifications that you can make from the **Context** menu of the filter and which add polish and specificity to your dashboard, are as follows:

- Applying **Quick Filters** to worksheets allows authors to select the sheets to which the filter applies, where all the worksheets are using the same data source

- Formatting **Quick Filters** gives authors control over the font faces and styles used in the title and the list of values

- Customizing **Quick Filters** can remove or show the **All** option

- Showing, hiding, or editing the title can prompt users to take action

- Changing the display of the filter, whether it's radio buttons for short lists or multiple value drop-down lists for longer lists, enables authors to select the best format for both the data and consumers

- Showing relevant values, also known as cascading filters, is a great way of simplifying the filtering experience for consumers so that selections in a filter determine the contents of another

Formatting the filter box so that it has a floating or fixed width is another feature that gives authors control over the appearance of worksheet.

Moving the Quick Filter

The quick filter is still present in the vertical container to the right that Tableau Public automatically created for it. If you have several **Quick Filters** and **Legends**, it's perfectly fine to leave them in the container. Just be conscious of how you are using *white space*.

We plan to add a parameter later, but for now, we want to put this quick filter next to the title for the dashboard. The title is in a horizontal layout container already, which means that we can put other objects next to it.

1. In order to drag the **Quick Filter** into the same horizontal layout container with the title, select the **Quick Filter** by clicking on it.

2. Hover the pointer of the mouse over the center of the **Quick Filter**. You can click on the area with the white hashes, which is outlined by a red box, as shown in the following screenshot:

3. When you get the white crosshair mouse, click on it and drag the filter into the horizontal layout container shared by the title for the dashboard. You will know that it is in the right place when the right border of the container has a shadow, as shown in following screenshot, and emphasized by the red box, as shown in the preceding screenshot:

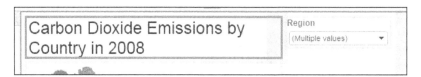

4. Note that the **Vertical Layout Container** has disappeared on its own.

5. Change the filter values and see how the dashboard display changes.

In the following screenshot, the changes that we made to the **Region** filter were applied to the other worksheets to which we had applied the filter (which included all the worksheets on the dashboard):

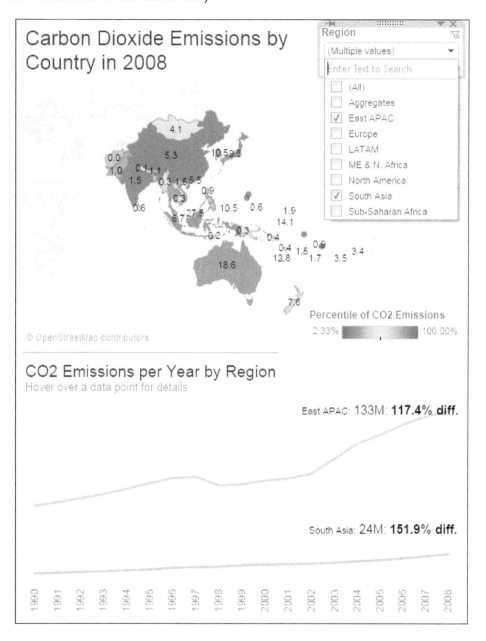

Filtering across Data sources with parameters

Although the **Quick Filter** can be applied to the worksheets using its data source, it can't be applied to worksheets using other data sources. In order to filter multiple data sources with one list, we need either a reference table or a parameter.

While you can use parameters for many important functions, such as determining which worksheets appear on a dashboard, there are some disadvantages:

- Only one value can be selected at a time
- The list of values for static parameters, and they need to be maintained manually
- Parameters are related to the **Data** sources through the use of calculated fields, which can be complicated to maintain

In the next few exercises, we will accomplish several tasks via the creation of three parameters. We will create parameters that integrate with calculated fields to:

- Show a single year in one graph and a range of years in another graph
- Filter across multiple **Data** sources
- Allow users to select the detail that needs to be explored

There are a couple of key concepts about parameters that you should keep in mind as we go through this chapter:

1. Each parameter is unique to a workbook and not to a **Data** source.

 You can copy parameters from one workbook to another without issues because they are not dependent on data sources.

2. Each parameter has a name, data type, format, and value.

The first task that we need to accomplish by creating a parameter involves using a **Year** filter for the dashboard. Two of the visualizations, namely the map and the line graph, have a **Time** dimension. We want to create a filter from which users can select the year. When a user selects a year, the map will be filtered to show only that year, but the line graph will show five years on either side of dashboard.

In order to build this functionality, we will need to create three new elements. The descriptions of these elements are as follows:

- The `Year` parameter: This has a data type of `INT` and a range of values

- A **Calculated** field: This will be used as a filter on the map so that the year that we are showing in the map matches the selection from the parameter

- A **Second Calculated** field: This will be used as a filter on the line graph that allows five years on either side of the selected year

Let's create a **Year** parameter by performing the following steps:

1. Create a new parameter by clicking on the **Context** menu in the **Data** window or by right-clicking on the white space in the **Data** window and selecting **Create Parameter**.

2. Give the parameter a name that represents what you want users to do. Ensure that the name does not match any of the field names in your data source. Otherwise, this will cause issues.

3. We named our parameter **Select a Year**.

4. Select the data type of the parameter. When you are using a parameter in a **Calculated** field, you can convert it to different types. We want ours to be a string because **Year** is a string field in the **Data** source.

5. If we had selected a different type of field, we would have had to format it accordingly. One of the reasons why we left it as a string is that we won't be running the risk of displaying improperly formatted values.

6. Select **List** for the **Allowable values**, as shown in the following screenshot:

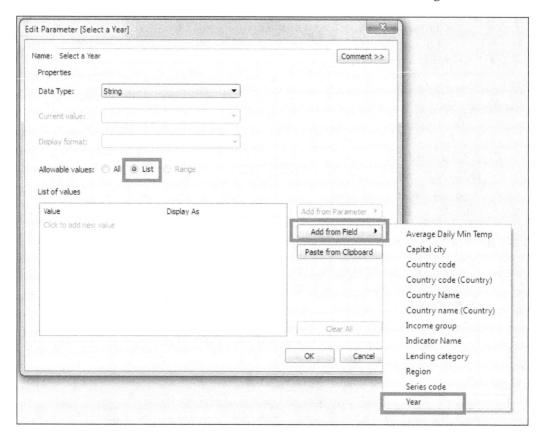

7. You have the option of populating the list from your own list of values or from one of the fields in either of the two **Data** sources.

 The parameter lists are not updated dynamically from the fields.

8. Click on the **Add from Field** button and then select the **Year** field.

 You only see fields of the same data type as a parameter. If there were multiple data sources in the workbook, you would see the others as well.

9. Now, the list of values matches that of the source field from the **Data source**. But from the bottom of the list of values, remove the years after *2008* because the data collected for these years is not complete. You can do this by hovering the pointer of the mouse over the row and clicking on the **X** that appears to the right of a values, as shown in the following screenshot:

10. Also, keep in mind that Tableau is case-sensitive, which means that the contents of the parameter list must match the contents of the field list.

11. The final parameter has been set to a default of **2008**, as shown in the following screenshot:

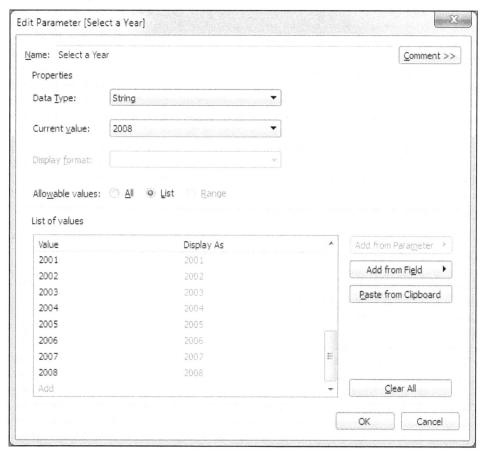

12. Now, click on **OK**.

Using parameters as Filters

You now have a parameter that you can use to filter certain fields. In the following steps, we will create two new calculated fields — the first will filter the map to show only the **Year** selected, and the second will filter the line graph to show five years on either side of the **Year** selected:

1. On the map's worksheet, the first task is to create a new **Calculated** field called **Map Year Filter**.

2. The formula is simple; it is Boolean expression, and it tells Tableau Public that we want only the **Year** values that map the parameter selection, as shown in the following screenshot:

3. Remove the filter on the **Year** field, which is a blue pill on the **Filters** shelf.

4. Drag this **new** field from the **Dimensions** pane to the **Filters** shelf and select **True**.

5. Apply this filter to the **Region KPIs** worksheet.

6. Right-click on the **Select a Year** parameter, which is below the **Measures** in the **Data** window, and show the parameter control.

Next, we will create a **Calculated** field for the line graph that filters to show five years on either side of the parameter.

Open the worksheet with the line graph, perform the following steps to create a **Calculated** field:

1. Create a new **Calculated** field and name it **Year +-**.

2. In the following screenshot, we used a formula to convert both the parameter and field to integers so that we can perform mathematical functions on them. If you want to, you can create another parameter to allow users to input the number of years that they would like to see on either side and then replace the instances of 5 with it:

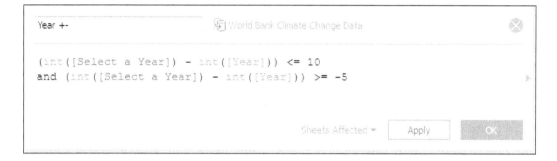

3. Click on **OK**.

4. Add this new field from the **Measures** pane to the **Filters** shelf and select **True**.

   Now, we will return to the dashboard and continue making modifications. The first thing that we want to do is show the parameter.

5. From the **Context** menu for the map worksheets on the dashboard, click on **Parameters** and then click on the **Select a Year** parameter.

6. The parameter automatically appears next to the **Region Quick Filter**. It would look better if the two were stacked. So, add a **Vertical Layout Container** to the **Horizontal Container** with the header and filter, and then move the **Filter** and parameter into it.

7. Add an extra 100 pixels of height to the dashboard.

Modifying titles

We can modify the header of the dashboard to reflect the selection of a year as well. The title is hard-coded to identify CO2 emissions from 2008, but if we select a different year from the parameter, such as 2005, then we need to ensure that the title is updated.

We can add the parameter to the title by performing the following steps:

1. Double-click on the title.

2. Delete the hard-coded reference to **2008**.

3. Click on the **Insert** button.

4. Select the parameter that we want to add, as shown in the following screenshot:

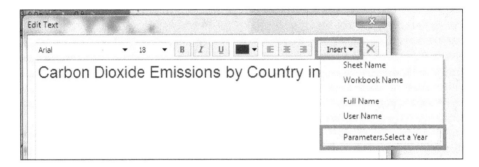

Since this title is actually a text object on the dashboard and not a specific worksheet title, we cannot insert field values, but we can insert parameters. It's always a good idea to make it very obvious what your user is looking at so that they don't run the risk of assuming the wrong thing.

Filtering across multiple Data sources with parameters

Parameters are useful when you need to filter across multiple **Data** sources. We loaded a second Data source, which contains data related to the foreign aid given to various countries by the United States since 1948, into our workbook. There are several fields that have the same name, namely **Region**, **Country**, and **Year**. We created a simple bar graph that shows the total aid in dollars over a period of time on a new worksheet called **Aid Graph**.

We created a filtering field for this visualization that shows only the **Year** selected in the parameter and the 30 years before it. In this case, we're using the same **Year** parameter that is being displayed on the dashboard. We also created a **Calculated** field, referencing it in the **Foreign Aid Data** sources, and then added it to the **Filters** shelf, just like we did on the map and line graph, as shown in the following screenshot:

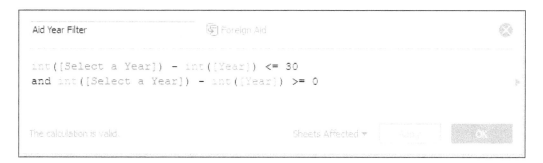

We will create a new parameter that allows users to select the visualization that they wish to view in the third of the last workbook, which is where we show granular details. Currently, the **Region KPIs** worksheet is displayed there, but we would like to provide users with the option of deciding what to see in the workbook.

In order to do this, we will create a new parameter. Then, we will create a **Calculated** field that can be used as a filter in each **Data** sources.

Let's create a parameter by performing the following steps:

1. Create a new parameter and name it **Select Granular Details**, which is a string with a list of two possible values.

2. We have capitalized these two values carefully because Tableau Public is case-sensitive. We will include following values in a **Calculated** field in the next step. The values are shown in the following screenshot:

    ° **Region Emissions KPIs**

○ **Foreign Aid Dollars**

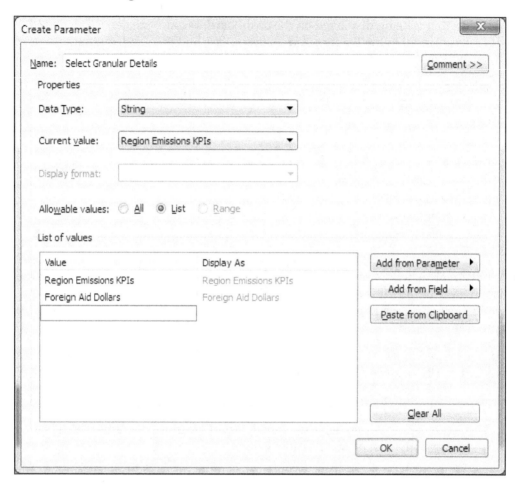

3. Click on **OK**.

4. In the **Foreign Aid Data** source, create a new **Calculated** field. This field will use a case statement to create a string output based on the parameter's selection.

5. We named the field **Granular Selection Filter**, and we wrote a case statement that states that if **Region Emission KPIs** is selected, then the output is **emissions**; otherwise, it's **Foreign Aid**. If there were more than two values, then we would have another condition, as shown in the following screenshot:

6. Click on **OK**.

7. Add the **Granular Selection Filter** field to the **Filters** shelf of the **Aid Graph**.

8. The string value that corresponds to the current value of the parameter will be displayed in the **Filter** list. Select it and click on **OK**.

9. Select **Show Parameter Control** and select the other value. The graph should now appear.

10. Right-click on the **Granular Selection Filter** field and select **Copy**. We'll paste it exactly as is into the **Data** source that contains the other worksheet on that we want to work.

11. Go to the **Region KPIs** worksheet. In the **Dimensions** pane, right-click on it and select **Paste**. The field that you just created now exists in both the data sources.

12. Right-click on the **Select Granular Details** parameter and select **Show Parameter Control**. You have to do this on every sheet on which you'd like to see it.

13. Ensure that the value selected corresponds to the graph that you're viewing.

14. Drag the **Granular Selection Filter** field to the **Filters** shelf and select **emissions**, as shown in the following screenshot:

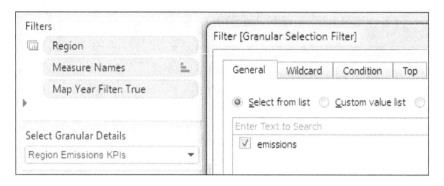

15. Click on **OK**.

Now, we can go back to the dashboard and add the parameter and the new visualization.

We will stack the visualizations into a **Vertical Layout Container**, and we will hide the title of each so that no space is unnecessarily taken up when it isn't selected. Then, we will modify the space in the title names of account for the parameter selection by performing the following steps:

1. First, add a **Vertical Layout Container** to the bottom of the dashboard and add the **Region KPIs** and **Aid Dollars** worksheets into it.

2. When you do this, the the parameters that appear in these worksheets will be added to the container with the filter and parameter at the top.

3. Add a **Horizontal Layout Container** within the new **Vertical Layout Container**.

4. Change the title for this area so that it is not floating and drag it into the **Horizontal Layout Container**.

5. Drag the parameter from the top into the **Horizontal Layout Container** as well.

6. Modify the text for the title so that it represents exactly what we can see in the following screenshot:

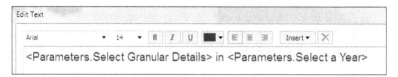

7. Right-click on the title for the **Aid Graph** and hide it.

8. Note that when you change the parameter value, both the title and the visualization change.

9. If you're having issues with the visualizations changing properly, the first thing that you need to check is the capitalization and spelling in the case statement. It's fairly common to make mistakes when writing these statements. Therefore, you should check them to ensure that you have written conditions that match the parameter values perfectly:

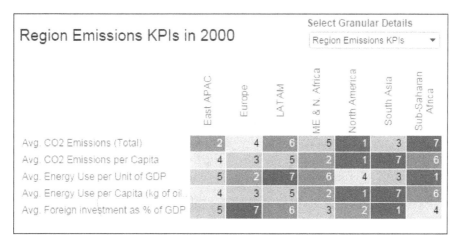

Actions

There are three types of actions that you can create on dashboards to add context, relevance, and specificity for your users, namely **Highlight**, **Filter**, and **URL**.

The following list describes these three action types:

*   **Highlight actions**: These draw attention to marks related to selections in a visualization without filtering other data points on the visualization.

*   **Filter actions**: These show only the selections made in a source visualization. They can be set up in such a way that when the marks in the source sheet are deselected, the target visualization shows all the marks, shows the filtered marks, or excludes all the marks.

*   **URL actions**: These allow you to link to generic URLs or dynamically add selected field values into URLs to link Tableau Public dashboards to external resources. As an example, we'll show you how to insert a country's name into a URL to directly link to its page on the World Bank's website.

Highlight actions do not filter visualizations, they only draw attention. URL actions link to external websites. Therefore, they don't have implications in dashboard performance. Filter actions have the following benefits:

- They are comparably faster than **Quick Filters**. Tableau Public does not need to scan the contents of a field before they can function.

- They can filter across multiple data sources without setting up join conditions ahead of time, which will be explored later on in this chapter.

- They allow users to select and view only the granular data that's important to them.

Actions have a disadvantage. For a novice author, maintaining actions can be time-consuming. The actions that we will use in the following exercises are designed to allow users to filter all the visualizations to show only the countries that interest them. Then, we'll add a URL action that links to the World Bank's web page for every individual country.

In the dashboard, we have a map that shows every country for which we have data. We have filters for time and other geographic dimensions. Therefore, we have an opportunity to do the following two things with the countries:

- We can create a filter action that limits the selections in other worksheets to the countries that a user has selected

- We can create a URL action that allows a user to see the World Bank's web page for each country

The first task involves creating the **Filter** action that runs when a user selects a country from the map by performing the following steps:

1. On the dashboard, click on the **Dashboard** menu and then click on **Actions**.

2. Click on **Add Action**, and then click on **Filter**.

3. Name the action **Filter on Country**. It's important to name actions according to their functions so that when you are editing and testing them later, you know where to look first.

The following attributes need to be selected:

- ° The source of the action, that is, the visualization from which it originates
- ° The action that triggers it
- ° The target sheets
- ° The fields on which you need to run an action

4. For the source, select the **CO2 Emissions per Capita** sheet on the dashboard.

5. Run the action on **Select**. Thus, when a user clicks on a field, its correlated marks in the target sheet will be highlighted. The alternatives are running the action when a user hovers over a mark or showing a hyperlink to run the action when a user rolls over **Menu**. We will use **Menu** for the URL action.

6. Target all the sheets on the dashboard except for the **CO2 Emissions per Capita** sheet.

7. Leave the default values to clear the selection. If we had worksheets that we wanted to hide until someone made a selection, which is a great option, we would have excluded all values.

8. Next, we need to establish the fields on which we want the filter to run. When a user clicks on a country, the value for that country is used as a filter for other sheets.

   In case you left the setting on **All Fields**, then all the attributes of the mark will be passed as a filter, but only to fields of the same name. For example, if you were going through the values for region and country, then each sheet with those exact same field names and values would be filtered.

   Since we have a secondary Data source, specifically the **Foreign Aid** data, that does not have the same field name for each country (though the values are the same) and has completely different values for the **Region** field, we need to establish targets.

9. Click on the **Selected Fields** radio button. Though using specific fields can potentially cause the dashboard to run more slowly with very large data sets, we need to establish the fields on which we need to join the **Data** sources.

10. Click on **Add Filter**. Since the action is running from a sheet using the World Bank Climate Change data source, it is the default for the source. Pick the **Country Name** field. Then, for the target, select the same **Data** source and field. This tells Tableau Public that for sheets that use this **Data** source, use the **Country Name** field as both the source and target. Check out the result in the following screenshot:

11. Click on **OK**.

12. Click on **Add filter** again. Select the **Country Name** field again as the source, but for the target, select the **Foreign Aid Data** sources and then select the **Country** field as the target.

13. Click on **OK**.

14. Check out the action dialog box, which is now complete, as shown in the following screenshot:

15. You can edit and remove the target filters as required.

16. Click on **OK**.

17. Back on the dashboard, modify the title above the map so that it says **Click on a country to filter** as sub-text in 12 pt Arial, as shown in the following screenshot:

18. Draw a box around an area of interest on the map. Check out the changes made to other visualizations in the dashboard, as shown in the following screenshot:

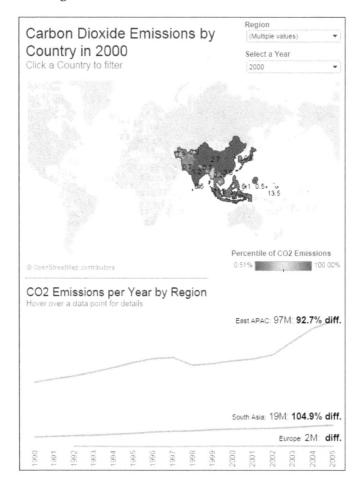

If we were to add additional visualizations to this dashboard, they automatically would become targets of the actions that we have established. For this reason, it's wise to wait until you have added everything of interest to the dashboard before creating actions.

URL actions

URL actions are useful when you wish to connect users to external and auxiliary data, whether it's from a public interest website or to a related dashboard. In the next example, you will create a URL action that allows users to view the World Bank's website for each country.

The URL action that you will build will append the name of each country as a variable to the root URL for the World Bank's well-organized websites.

The root URL is `http://www.worldbank.org/en/country/`, and in order to get to a specific country, such as South Africa, you need to add the country name so that it looks like `http://www.worldbank.org/en/country/southafrica`.

There is one issue — there are no spaces in the country names in the URL and yet, there are spaces in the country names in the data source.

So first, create a **Calculated** field that removes the space in each **Country** name so that we can append it as a token to the URL.

To remove and append the **Country** name with a URL, perform the following steps:

1. Go to the **Map** worksheet and create a **Calculated** field in the **Data** source.

2. Name the **Calculated** field **CountryName**.

3. In the **Calculated** field, write a formula that uses the **Replace** function to remove the spaces in the **Country** field, as shown in the following screenshot. You need to remove the spaces because in the URL, for each country, there are no spaces. Different websites treat spaces differently, depending on their information management strategy and encoding; others might replace spaces with %20:

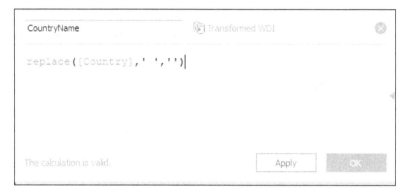

4. Add this field to the **Detail** shelf on the sheet.

5. Note that the percentile table calculation that generates the color for each country no longer seems to work because it was addressing only by **Country Name** before and it wasn't partitioned (or grouped) by any other values.

6. In order to solve this problem, click on the context menu for the field on the **Color** shelf.

7. Click on **Edit Table Calculation**.

8. In the **Running Along** drop-down list, select **Advanced**.

9. Click on the **CountryName** field in the **Partitioning** pane and then click on the **right arrow** to move it over to the **Addressing** pane, as shown in the following screenshot:

10. Click on **OK**.

11. Go back to the dashboard. From the **Dashboard** menu, select **Actions**.

12. Create a new **URL** action.

13. Name the **URL** action with the text that will prompt users to perform a specific action, such as **Click on Here for Country Data**. You want the result to be very obvious because this action will be a **Menu** action, which means that the name of the action will appear as a call to an action at the bottom of the tooltip, as shown in the following screenshot:

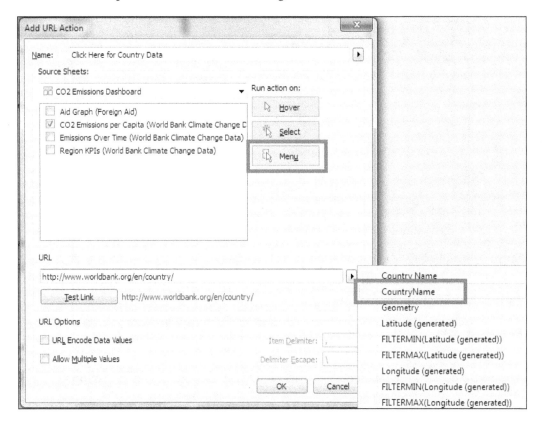

14. Select **CO2 Emissions per Capita** as the source sheet.

15. Run the action on **Menu**.

16. Paste the root URL for the World Bank into the **URL** box. Then, click on the arrow to the right of the URL box to see a list of fields that you can insert as a variable token.

17. Select **CountryName** from the list.

18. Test the link by clicking on the **Test Link** button and, if it works, click on **OK**.

19. Click on **OK** again.

20. Click on a **Country** (**China**, for instance) and check out the call to action at the bottom of the tooltip, which you unfortunately cannot format differently, that guides the user to click on **Click Here For Country Data** to see country data in detail, as shown in the following screenshot:

Summary

In this chapter, you learned how to use **Quick Filters** and **Dashboard Actions** not only to provide greater interactivity for users, but also to provide an efficient user experience. You also learned how to create parameters to filter single and multiple data sources, as well as to govern the visualizations that appear on a dashboard. You studied the implications in dashboard performance when using different methods of filtering, all of which will give your dashboard users a greater control over the data points that they choose to explore. It will also create a more personal, relevant, and compelling data story.

In the next chapter, we will discuss publishing your work, which is the last step to creating and distributing your work in Tableau Public.

Publishing Your Work

9

Publishing your work on the Cloud with Tableau Public is the core of the application's value proposition, and with Tableau Public 9.0, it's easy to save and share your work. There's more to publishing than just locating a file on a web server. Publishing your work and managing your author and workbook profiles is the best way of helping people browse through your visualizations to create more personal connections with your data story. When people are personally connected and relate to you and your work as a human, they are more likely to share it.

In this section, we'll discuss the following topics:

- Saving your work and logging in to Tableau Public
- Opening your work from the Cloud on your computer
- Managing your profile
- Viewing your work online
- Sharing your work with others
- Managing workbook details

Saving your work and logging in to Tableau Public

Since Tableau Public does not have autosave, we recommend that you save to the Cloud frequently.

In the **File** menu, you can save your work by clicking on the **Save to Tableau Public...** option, which is present halfway down the list. Alternatively, you can use the same shortcut keys that other applications use (*Ctrl + S*):

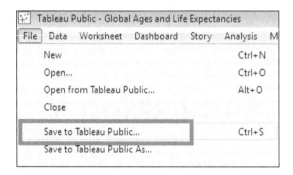

Once you opt to save your work, you will need to sign in to Tableau Public. Use the same credentials that you used to download Tableau Public initially, as shown in the following screenshot:

Once you have signed in successfully, you can save your work on the Cloud with a name of your choice.

Giving your analysis a good name is critical. First, it's the title that other people will see when they look at your profile on Tableau Public. Also, the name of the workbook and individual dashboards are a part of the URL of the file. It's important not to include special characters, such as ampersands, question marks, and periods, because they can disrupt the format of the URL. It's fine to include numbers though. You can name your workbook as shown in following screenshot:

If you want to replace an existing file with the one that you have opened now, you can perform the following steps:

1. Select **Save to Tableau Public As...** from the **File** menu.
2. From the dropdown list of the existing files, select the file that you want to overwrite.
3. Click on **Save**.

Opening work from the Cloud

If you'd like to make changes to your work, you need to perform the following steps:

1. Open Tableau Public on your computer.
2. Navigate to the Tableau Public home screen.
3. Click on the orange **Open from Tableau Public** link on the upper-right corner.
4. Log in with the credentials that you used to create your account, which was covered previously in this chapter.

5. Select the workbook that you want to open, as shown in the following screenshot:

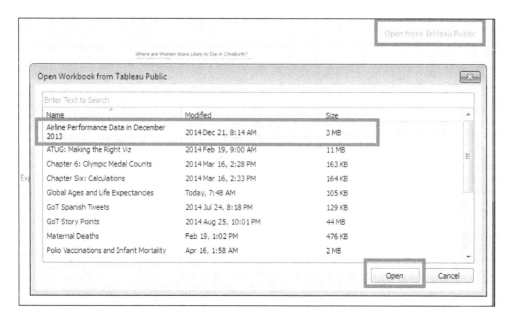

Managing your profile

In order to view your work on the Internet, you need to go to Tableau Public's website and log in. From there, you can share your work with others, download it, promote it on social media, and also delete it, should the need arise.

1. On your Internet browser, go to `http://public.tableau.com`.

2. In the upper-right corner, click on **SIGN IN**, as shown in the following screenshot:

3. Enter the credentials that you used to create your account.

The destination page that you reach when you login is your profile page. This is where all the dashboards that you created are organized. It is also a great place for you to add information about yourself. You can add links to your Twitter and LinkedIn accounts as well as the blogs or corporate websites that you would like to promote. (It's important to make sure that you have the permission before presenting URLs of organizations of which you're not an associate).

My profile is shown in the following screenshot:

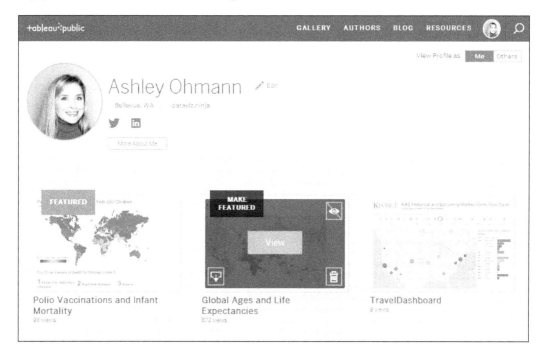

You can change settings related to specific workbooks by hovering over them. As your catalogue of work grows, you should curate it by being selective about which visualizations are featured.

When you roll over any of the thumbnails for a specific visualization, you can do several things; the following are a few of these things:

1. First, on the upper right-hand side, select **MAKE FEATURED** to make it the featured visualization, which will move it to the upper-most left spot on your profile.

2. On the upper right-hand side, you can change the **visibility** settings.

This does not prevent people from searching for your visualization, and you also cannot control the permissions regarding who can see it.

3. On the lower right-hand side, you can delete workbooks. This is a good idea in case you would like to delete the earlier iterations of a finished product, but keep in mind that you don't have the file saved on your computer. So, be judicious when using the **Delete** button.

4. On the lower right-hand side, you will see the workbook name and the number of views that it has.

 A view is each visit to the page by a browser. So, if you go to the workbook five times a day, that's five views. Note that this is not the number of unique visitors.

5. Then, above the workbook name, you can download the workbook.

6. At the center, you have the **View** option in case you wish to view the workbook, as shown in the following screenshot:

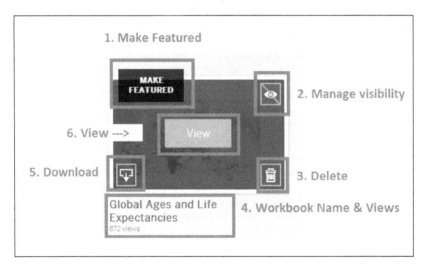

Viewing your work online

When you view a workbook, you have several options at the top of the page.

From left to right, you can perform the following tasks:

- Go back to your profile
- Keep on clicking and go through other workbooks
- Edit the details of the workbook that you have opened
- Download the workbook, as shown in following screenshot:

If you scroll to the bottom of the dashboard, you have the following additional controls:

1. **Undo**, **Redo**, and **Reset** any changes that you have made. For instance, if you have filtered your workbook and want to revert to the original state, just click on the **Reset** button.

2. Share your work in the following way:

   ○ You can share the **Current View**, which includes filter conditions or parameter selections. Alternatively, you can share the **Original View**.

   ○ **Embed Code** allows you to copy an automatically generated block of HTML that you can use to embed the visualization on another web page. This code identifies the viewing attributes of your workbook, such as whether the toolbars are showing, the height and width of the workbook, and its title. For instance, if you want to embed the dashboard on your blog, you just need to copy the code. Then, in the management console of your blog, either insert the code directly, or create and then insert a snippet of HTML. If you have the capabilities, you can utilize Tableau Public's JavaScript API to embed the visualization in a webpage and then allow users to interact with it through web objects that are native to the source page. That's beyond the scope of this chapter though.

- ° To share the workbook with others, you can copy the link as well as simply highlight and copy the text and then paste it into a new e-mail in the tool of your choice.

- ° You also can share the workbook on Twitter or Facebook by clicking on either of the two buttons that open a pop-up on your browser and ask you to enter your credentials for the medium that you have selected, as shown in the following screenshot:

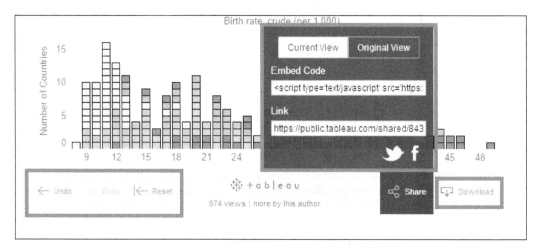

When you share your work via Twitter or Facebook, the description that you enter in the workbook details will automatically be populated, as shown in following screenshot. You can always change the description and remove the `via @tableau` string, but it's much more likely that people will find your tweet if you leave that tag:

Managing workbook details

The information that you provide about your work is important; it's the first part of the story that you're telling that people will take in. The workbook name is important, and so is the description and additional links.

If you scroll down the page past the **Share** button, you will see the workbook title as well as a link that allows you to edit the workbook details, as shown in the following screenshot:

Now, perform the following steps:

1. Make sure that the Title of the workbook says exactly what you want to appear on your profile.

2. Add a **Permalink**. Though this isn't required, it's a good way to promote your blog.

3. Enter a good description in the Description. This is the text that search engines will index. Also, if you want people to tweet your work, then you need to enter a description that's 82 characters or less, because the link that Tableau generates for your workbook will consist of 62 characters.

4. Toolbar Settings typically include control buttons as well as a link to your author profile. If you choose not to include those, then the HTML generated to embed your work will be adjusted.

5. An additional setting that you can adjust is **Show workbook sheets as tabs**. If all the worksheets in your workbook are on your dashboard and you have hidden them, then you don't need to select this.

6. Save your work by clicking on **Save**, as shown in the following screenshot:

Summary

In this chapter, you learned how to save your work to the Web as well as how to manage your Tableau Public profile and workbook attributes to help you showcase and share your work effectively.

This chapter concludes our discussion of creating data stories in Tableau Public. Throughout the previous eight chapters, we progressed through the basics of the tool, data analysis, and different visualization types, all the way through sophisticated calculations and parameters. You now have the skills to create rich, relevant, and compelling data stories from which others can learn, get inspired, and find the information that they need to improve their own lives and those of others.

Index

MAKETIME function, using 91
STR function, using 91
types of functions
  aggregate calculations 87
  date 87
  logical 87
  number 86
  strings 86
  table calculation 87
  type conversion 87
  user functions 87

U

URL actions
  defining 175-178
user interface, Tableau Public
  about 17, 18
  Analytics pane 22, 23
  canvas 25-27
  cards and shelves 19
  Column/Row shelves 25-27
  Columns and Rows shelves 27-29
  data pane 20, 21
  Data Source 20
  menus 23-25
  Sheet tabs 19
  ShowMe card 19
  side bar 20
  Start button 20
  Status bar 19
  toolbar 19-25
  View or Visualization 19
  visual cues 22
  workbook 19

V

virtualization
  issues 127
visual clues
  perception 56, 57
visual cues
  Abc field 22
  about 22
  calendar icon 22
  globe icon 22
  paper clip icon 22
  # sign 22
  Venn diagram icon 22
visualization
  bar charts 70-72
  geographic maps 73, 74
  groups 78-80
  line graphs 65, 66
  pie charts 76, 77
  scatter plots 74-76
  sets 78-80
  tables 66-69
  types 64
VIZ OF THE DAY gallery 14

W

WINDOW function
  versus RUNNING function 123
work
  online viewing 186-188
  opening, from Cloud 183
  saving, to Cloud 181-183
workbook details
  managing 189
workspace control tabs, Tableau Public
  about 32, 33
  Show Filmstrip 33
  Show Sheet Sorter 33
  Show Tabs 33

Thank you for buying
Creating Data Stories
with Tableau Public

About Packt Publishing

Packt, pronounced 'packed', published its first book, *Mastering phpMyAdmin for Effective MySQL Management*, in April 2004, and subsequently continued to specialize in publishing highly focused books on specific technologies and solutions.

Our books and publications share the experiences of your fellow IT professionals in adapting and customizing today's systems, applications, and frameworks. Our solution-based books give you the knowledge and power to customize the software and technologies you're using to get the job done. Packt books are more specific and less general than the IT books you have seen in the past. Our unique business model allows us to bring you more focused information, giving you more of what you need to know, and less of what you don't.

Packt is a modern yet unique publishing company that focuses on producing quality, cutting-edge books for communities of developers, administrators, and newbies alike. For more information, please visit our website at www.packtpub.com.

Writing for Packt

We welcome all inquiries from people who are interested in authoring. Book proposals should be sent to author@packtpub.com. If your book idea is still at an early stage and you would like to discuss it first before writing a formal book proposal, then please contact us; one of our commissioning editors will get in touch with you.

We're not just looking for published authors; if you have strong technical skills but no writing experience, our experienced editors can help you develop a writing career, or simply get some additional reward for your expertise.

Learning QlikView Data Visualization

ISBN: 978-1-78217-989-4   Paperback: 156 pages

Visualize and analyze data with the most intuitive business intelligence tool, QlikView

1. Explore the basics of data discovery with QlikView.

2. Perform rank, trend, multivariate, distribution, correlation, geographical, and what-if analysis.

3. Deploy data visualization best practices for bar, line, scatterplot, heat map, tables, histogram, box plot, and geographical charts.

Python Data Visualization Cookbook

ISBN: 978-1-78216-336-7   Paperback: 280 pages

Over 60 recipes that will enable you to learn how to create attractive visualizations using Python's most popular libraries

1. Learn how to set up an optimal Python environment for data visualization.

2. Understand the topics such as importing data for visualization and formatting data for visualization.

3. Understand the underlying data and how to use the right visualizations.

Please check **www.PacktPub.com** for information on our titles

Tableau Data Visualization Cookbook

ISBN: 978-1-84968-978-6   Paperback: 172 pages

Over 70 recipes for creating visual stories with your data using Tableau

1. Quickly create impressive and effective graphics which would usually take hours in other tools.

2. Lots of illustrations to keep you on track.

3. Includes examples that apply to a general audience.

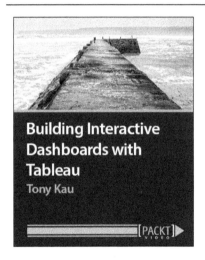

Building Interactive Dashboards with Tableau [Video]

ISBN: 978-1-78217-730-2   Duration: 04:31 hours

Create a variety of fully interactive and actionable Tableau dashboards that will inform and impress your audience!

1. Increase your value to an organization by turning existing data into valuable, engaging business intelligence.

2. Master the dashboard planning process by knowing which charts to use and how to create a cohesive flow for your audience.

3. Includes best practices and efficient techniques to walk you through the creation of five progressively engaging dashboards.

Please check **www.PacktPub.com** for information on our titles

www.ingramcontent.com/pod-product-compliance
Lightning Source LLC
LaVergne TN
LVHW062315060326
832902LV00013B/2227